The Militarisation of Space

The Militarisation of Space

Edited by

Stephen Kirby
Director, Centre of Defence Studies
University of Hull

and

Gordon Robson
Senior Lecturer in International Politics
Sunderland Polytechnic

WHEATSHEAF BOOKS, SUSSEX

LYNNE RIENNER PUBLISHERS, BOULDER

First published in Great Britain in 1987 by
WHEATSHEAF BOOKS LTD
A MEMBER OF THE HARVESTER PRESS PUBLISHING GROUP
Publisher: John Spiers
16 Ship Street, Brighton, Sussex

and in the United States of America by
LYNNE RIENNER PUBLISHERS, INC.
948 North Street, Boulder, Colorado 80302

British Library Cataloguing in Publication Data
The Militarisation of space
1. Space warfare
I. Kirby, Stephen II. Robson, Gordon
358′.8 UG1530

ISBN 0-7450-0346-X
ISBN 0-7450-0412-1 Pbk

Library of Congress Cataloging-in-Publication Data
The Militarisation of space edited by Stephen Kirby and Gordon Robson
p. cm.
Includes bibliographies and index
ISBN 1-55587-103-8 1. Astronautics, Military
I. Kirby, Stephen. II. Robson,
Gordon
UG1520. M55 1988 87-25064
358′8--dc19 CIP

Typeset in 11pt Times by
C. R. Barber & Partners (Highlands) Ltd.,
Fort William, Scotland.

Printed in Great Britain by
Billing & Sons Ltd., Worcester.

Contents

Notes on Contributors

IAN BELLANY, a physicist by training, is Director of the Centre for the study of Arms Control and International Security at the University of Lancaster. He has written extensively on arms control and defence issues and is editor of the journal *Arms Control.*

MICHAEL CLARKE is lecturer in International Politics at the University of Newcastle upon Tyne, specialising in British foreign policy and defence politics. Recent work concentrates on non-nuclear defence policies in NATO and on defence relations between the United States and Western Europe.

Group Captain TIMOTHY GARDEN is a serving officer who currently commands Royal Air Force Odiham operating Chinook and Puma helicopters, and was recently Director of Defence Studies for the RAF. He writes on air power, defence and nuclear issues and is a Council Member of the Royal United Services Institute.

STEPHEN KIRBY is lecturer in International Politics, and Director of the Centre for Defence Studies, at the University of Hull. He has taught in France, South Korea and in military establishments throughout the United Kingdom. He specialises in the politics of Western defence.

Squadron Leader ROBERT LAPRAIK is a graduate of Aeronautical Engineering and has served with the RAF both operationally and as an instructor. He was appointed to the

RAF Aerospace Briefing Team in 1984 and has lectured throughout the UK and overseas. He retired from the RAF in November 1984 to take up a General Manager post in the National Health Service.

TERRY McNEILL is a lecturer in Soviet Politics at the University of Hull who writes, teaches and broadcasts extensively on contemporary issues in Soviet society.

GORDON ROBSON is a senior lecturer in International Politics at Sunderland Polytechnic, specialising in the study of Western security. He has also taught at the University of Hull and for the Home Office at Dishforth Police Training Centre, and is currently on secondment to the Polytechnic's External Development Unit, furthering industrial contacts.

STEPHEN SHENFIELD is on the research staff of the Centre for Russian and East European Studies at the University of Birmingham. He has published articles on Soviet foreign and defence policy.

STEVE SMITH is senior lecturer in International Relations at the University of East Anglia, and has also taught at the State University of New York (Albany). He has written widely on nuclear weapons technology, arms control and foreign policy analysis and is editor of the Cambridge University Press series 'Studies in International Relations'.

Group Captain MARTEN VAN DER VEEN is a graduate of Engineering Science and has been a serving officer with the RAF since 1963 in operational, instructional and administrative roles. He has served as exchange officer with the French Air Force, spent several years at the Ministry of Defence and in May 1985 was appointed Director of Defence Studies for the Royal Air Force.

PHILIP WEBBER is a physicist who has conducted research at the Imperial College of Science and Technology, London, over a twelve-year period. He has specialised in the consequences and effects of nuclear war and nuclear accidents.

Current interests include the 'Star Wars' programme and conventional defence in Europe. He is currently a local government officer in South Yorkshire.

Abbreviations

ABM	anti-ballistic missile
ACP	airborne command post
AFB	Air Force Base
AFSATCOM	Air Force Satellite Communication System
ASAT	antisatellite
ATBM	anti-tactical ballistic missile
AWACS	airborne warning and control system
BMD	ballistic missile defence
BMEWS	ballistic missile early-warning system
C^3	command, control and communication
CCD	charge-coupled device
CDMA	code division multiple access
C^3I	command, control, communication and intelligence
CIA	Central Intelligence Agency
CINCLANT	Commander in Chief, Atlantic
CINCPAC	Commander in Chief, Pacific
DARPA	Defense Advanced Research Projects Agency
DF	deuterium fluoride
DOD	Department of Defense
DSAT	satellite defence
DSCS	Defence Satellite Communications System
DSP	Defense Support Program
DYNASOR	experimental one-man winged spacecraft
ECs	European Communities
ECCM	electronic counter-counter-measures

ECM	electronic counter-measures
EHF	extremely high frequency
ELINT	electronic intelligence
EML	electromagnetic launcher
EMP	electromagnetic pulse
ERIS	Exoatmospheric Re-entry Interception System
EUREKA	European Research Co-ordination Agency
EW	electronic warfare
FDMA	frequency division multiple access
FEL	free-electron laser
FET	field effect transistor
FLTSATCOM	Fleet Satellite Communications System
FOBS	Fractional Orbital Bombardment System
FOFA	follow-on force attack
GaAs	gallium-arsenide
GBL	ground-based laser
GEODSS	Ground-based Electro-Optical Deep Space Surveillance
GHz	gigahertz
GLONASS	Global Navigation System (USSR)
HEDI	High Endoatmospheric Defense Interceptor
HEL	high energy laser
HF	high frequency
HOE	Homing Overlay Experiment
ICBM	intercontinental ballistic missile
IDEA	Initiative for the Defence of European Aerospace
IDEACO	IDEA Co-ordination Office
IEEE	Institute of Electrical and Electronic Engineers
IISS	International Institute for Strategic Studies
IMPATT	impact avalanche transit time
INF	intermediate-range nuclear forces
IR	infrared
IRBM	intermediate-range ballistic missile
KEW	kinetic energy weapon
MAD	mutual assured destruction
MHV	miniature homing vehicle
MHz	megahertz

MILSTAR	Military Strategic–Tactical and Relay Satellite System
MIRACL	mid infrared advanced chemical laser
MIRV	multiple independently targetable re-entry vehicle
MIT	Massachusetts Institute of Technology
MMIC	monolithic microwave integrated circuit
MOL	manned orbiting laboratory
MSR	missile site radar
MSS	multispectral Scanner
NAVSTAR.GPS	Navigation and Satellite Timing and Ranging Global Positioning System
NCA	national command authority
NEACP	National Emergency Airborne Command Post
NNK	non-nuclear kill
NORAD	North American Aerospace Defense Command
NOSS	Naval Oceanographic Surveillance System
NSDD	National Security Decision Directive
NSDM	National Security Decision Memorandum
OMG	operational manoeuvre group
PAR	perimeter acquisition radar
PCM	pulse code modulation
PD	Presidential Directive
PMAL	prototype miniature air launched
RBV	return beam vidicom
R&D	Research and Development
RV	re-entry vehicle
SAC	Strategic Air Command
SAINT	satellite interceptor
SAM	surface-to-air missile
SAR	synthetic aperture radar
SBKKV	Space-based kinetic kill vehicle
SDI	Strategic Defense Initiative
SDIO	Strategic Defense Initiative Office
SDMA	space division multiple access
SDS	satellites data system
SGEMP	system-generated electromagnetic pulse
SHF	super high frequency

SIOP	Single Integrated Operational Plan
SLBM	submarine-launched ballistic missile
SRAM	short-range attack missile
STEW	space-to-earth weapons
TACAMO	submarine communications aircraft (It is not certain that this is an acronym but it is usually taken to stand for 'take charge and move out')
TDMA	time division multiple access
THOR tests	Tiered Hierachy Overlayed Research
TWT	travelling-wave tube
UHF	ultra-high frequency
WWMCCS	World-Wide Military Command and Control System

Introduction: States, Space and Security

Gordon Robson

Space has been essential to the calculus of superpower relationships for over 30 years, and about 75 per cent of satellites have some military application.[1] This process, as van der Veen illustrates, has contributed significantly to the current superpower strategic regime whilst also acting as a 'force multiplier' of terrestrial forces, qualitatively improving existing military systems through improvements in C^3I. The 'high frontier' has become, in consequence, critical to the structure of superpower stability—a requirement for defence, deterrence and reassurance. Space, despite variations in superpower doctrine, is now inexorably linked to their perceptions of both national and common security.

In the varied but subtle world of superpower strategic relationships, space has taken on a chameleon-like quality, able to adapt yet remain central, to the assumptions of diametrically opposing perspectives. On the one hand is the view that space can support strategies which enhance deterrence and stability, by providing some measure of protection to retaliatory forces, and this area is examined by Lapraik. This same view also regards space as essential for the maintenance of stability through surveillance, verification and communications. Satellites, by ensuring the passage of clear and unambiguous signals between the superpowers, have reduced mistrust and the danger of miscalculation, and improved early warning systems thereby reducing the fear of surprise attack. All of which have become enshrined in the arms control regime and absorbed into the notions of crisis management and confidence-building, whether in the superpowers' central relationship or in their regional spheres of influence and interest.

In contrast to this, President Reagan's Strategic Defense Initiative, announced in March 1983, has given hope to two groups which oppose the nuclear status quo and current defence theory. The first, 'visionaries', see an 'astradome-type' shield to be preferable to the continued captivity of deterrence and regard a superpower relationship based upon the threat and mutual fear of destruction, to be repugnant and alien. Deterrence *per se* is immoral, irrational and unstable. For this group the frontiers of new technology are to be harnessed for active defence rather than to support the stalemate of deterrence. High technology, when applied to space, offers a moral alternative based upon total defence from missile attack. This positive if utopian view of space is clearly portrayed in Reagan's vision of SDI and assumes that the theoretical laws of physics can be applied as a technical solution to a political problem. Clarke examines whether or not this may be counterproductive, particularly as improved ASAT capability could damage US security and force structures dependent as they are upon vulnerable space-based assets. However in its purest form SDI might also offer a radical solution to the nuclear dilemma, defence rather than deterrence. It has been argued that the programme has an ethical content as it may save lives, encourage the disarmament process, reduce the occurrence of war, and enhance the legitimacy of western defence policy. As Colin Gray has said, 'A technological mature SDI would enable the United States and its allies to defend their values against nuclear intimidation or even actual nuclear assault without, in the process, negating everything that they stand for'.[2]

A second group sees in the defensive capability of SDI a competitive advantage which requires not only militarising, but also 'weaponising' space. To many Americans the SDI encapsulated a new mood which touches 'all the right buttons like security, enterprise, strength and virtue' and which satisfies 'both wings of the nuclear weapon strategists, the ideologues who believed in US military superiority and those pragmatists who thought the Soviets were cheating at arms control.'[3] This has enabled a return to the classical view of world politics where state security is firmly a function of military power and superiority. ASATs, space battle-stations, and nuclear war-fighting options, have developed due to the marriage of classical strategic preference and exotic technology. Refinements in policy and developments in technology have weakened the nuclear stalemate, the great threat if paradoxical comfort of the late 1960s and 1970s. Central to this, undoubtedly, has been the growing militarisation of space and prior to that, of science and technology. This indicates that SDI has become 'the ultimate technological fix'[4] and suggests a deep structural commitment by both industry and thc military to technical salvation.

Thus President Reagan's speech has encouraged powerful lobbies to embrace the opportunities afforded by the new technology. They are often bitterly at odds over such fundamental issues as the pace of development and deployment, and the feasibility of total missile defence. As early as 1983 the Fletcher study indicated the long-term programme whilst the Hoffman panel defined an interim menu of options.[5] With the passage of time the pressures of both technology and political vicissitudes have indicated the acceptance of more limited goals but fundamental disagreements continue. For example Richard Perle, Assistant Secretary of Defense, identifies the first role of SDI as 'not the defense of the nation as a whole, not of every city and person in it, but the defense of America's capacity to retaliate'.[6] Whereas Weinberger reaffirmed, in July 1986, commitment to total defence and stated that 'it's not our missiles that we seek to protect but our people. . . . And we must never lose sight of that goal'.[7] This return to the concept of total defence, with a subsequent commitment to early testing despite the ABM Treaty, now represents the most hawkish view in the White House, and one likely to be at odds with the Democratic controlled Congress during the twilight of the Reagan Administration.

Thus to the strategic hawks of the Reagan Administration space provides an opportunity not only to regain the moral high ground but also to regain nuclear preeminence. They suggest that first-strike assets can be made invulnerable and that eventually space-based weapons may be used against terrestrial targets. Beyond that there is a political and diplomatic bonus; it may pressurise the Soviet economy and become a 'bargaining chip' for arms control restraint and even behavioural change. Ultimately it facilitates a preference for superiority, a unilateral solution to the universal problem of security, particularly when linked to a first-strike capability. The language has changed but not the message, superiority is still security.

Ironically, as the superpowers embrace the arming of space so the ideal of 'common security' may retreat,

> No matter how many weapons a nation adds to its arsenal, it cannot directly diminish its vulnerability. No known technology, provides, even potentially, a means for the effective and reliable defence of a people from nuclear attack. Thus, one central irony that must be faced is that no matter what unilateral choices a nation makes in pursuit of security, it will remain vulnerable to nuclear attack and thus ultimately insecure.[8]

Rationality in crises must assume some minimum of reflection, some calculation of risk and the likely costs and benefits of any particular action. In space-based defence there will be no time for even this minimum of human doubt as computer response alone will perform the calculations and must in the logic of instant reaction 'launch on doubt'. Indeed the sheer complexity of the computer systems, and the possibility of error in systems which cannot be tested in 'real time' nuclear war conditions, explains the scepticism of many from within the scientific community.[9] To specialists such as David Parnas and Herbert Lin, designing the necessary software will be impossible.[10] The set-back to the US space programme since the Challenger disaster in January 1986 has compounded the problem and added weight to those who doubt the feasibility of the new technology and viability of a full multi-layered BMD. Despite frequent claims of progress many within the scientific community still remain scornful, and doubt in particular if identification, tracking and communication systems can be viable, a debate examined by Webber.

Therefore it can be seen that the accelerating militarisation of space has forced attention upon key strategic issues; the effect upon the arms race, alliance relations, the arms control regime and overall superpower stability. The superpower decision to pursue research into space defence naturally contributes to the global arms competition and, as Shenfield examines, may have implications for terrestrial as well as space-strategic systems. Further, whether visionary or not, a total BMD system may make conventional war in Europe more acceptable if the threat of nuclear exchange had been removed by a defensive shield. To offset this possibility a conventional rearmament would seem a prerequisite.

The US embrace of strategic defence has produced contradictory challenges and opportunities for the Western Alliance and these are considered by Kirby. To the proponents of SDI Europe can benefit from strategic defence if it is linked to an anti-tactical ballistic missile system, and allies may also benefit from the technology transfer which could follow participation in research and development. Despite these possibilities US policy has fuelled transatlantic frictions on a broad range of issues, which include US-European coupling, the first use of nuclear weapons, the viability of British and French nuclear deterrents, and the allocation of defence responsibilities. Other issues include the degree of burden-sharing, and the nature of technology transfers within the Alliance, and implications such as the cohesion of Western societies, the role of European collaboration, and the evaluation of East–West relations, including in particular the role of arms control.[11]

Since 1983 superpower space intentions have both encouraged and weakened the arms control regime, and this complex relationship is analysed by Bellany. By 1986 it had become clear to the Soviet Union that the US research programme was progressing so rapidly that arms control could not be neglected and that through negotiation the US effort may be restrained, and McNeill examines the success of this policy. The optimism of the negotiating table was witnessed on the roads to Geneva and Reykjavik in 1986, although dashed by the superpower failure to compromise over SDI and by the US insistence on a liberal reinterpretation of the ABM Treaty. This in the opinion of the proponents of SDI is vital, not only because many past agreements have been flawed but because they 'reflect the views, ideals, and technologies of the time' . . .

and . . . 'they should not be viewed as sacrosanct'.[12] As SDI can not be isolated from the general context of other force modernisations, it can be seen to have held back agreements to reduce strategic and intermediate forces, made qualitative limitations difficult, and generally created an unsympathetic climate for progress in confidence and security building measures. Alternatively, to some within the US Administration, the pace of advance has provided a 'bargaining chip, which should be used to obtain a package of limitations on defensive and offensive forces.'[13] Amongst the most pressing limitations for both crisis stability and US security is that of an anti-satellite agreement and this is analysed by Smith.

Ultimately the concern of proponents and opponents of the militarisation of space is the issue of stability which is 'that condition which minimises the probability of nuclear war or the highly provocative behaviour that might lead to nuclear war'.[14] Over the past 20 years this has come to mean a condition of mutual assured vulnerability which a technological breakthrough such as a total or near total defence could so easily unbalance. The possibility of a unilateral Soviet deployment has been the rationale for US concern and its commitment to SDI. The opponents of that policy have catalogued a series of factors which induce destabilisation, including pre-emptive and first-strike scenarios, predelegation and miscalculation through technological errors. Nuclear defence or BMD as currently envisaged may provide the opportunity and pressure for a pre-emptive strike—the technology could be adapted to this role whilst the less developed superpower may perceive the need for restorative action. To many critics of current US policy, the Soviet Union is as likely to be provoked by the SDI proposals as it is to be deterred. Hence in a worst-case scenario there is no reason for the Soviet Union to accept the US argument that BMD represents the fine tuning of deterrence and a high-tech solution to the scourge of nuclear war. It is a threat.[15] On the other hand the mutual deployment of strategic defence may well stabilise nuclear deterrence by preventing uncertainty and tension, by countering conventional ballistic missiles, a possibility considered by Garden. Eventually one's perspective becomes a matter of faith.

This volume brings together essays written by specialists from both the academic and serving military communities of the United Kingdom. It addresses and analyses the fundamental questions about the militarisation of space and its impact upon the central East–West strategic relationship. In so doing it assesses the application of science and technology to nuclear doctrine. Each contribution, to some degree, offers a recognition and an awareness that although technology cannot be disinvented its application involves political decision. In this there can only be optimism for the militarisation of space, although inevitable, is not immutable and this volume helps to explore the limits of choice that still exist.

NOTES

1. B. Jasani and C. Lee *Countdown to Space War* (Taylor and Francis, London, 1984) p. 1.
2. C. Gray, *Strategic Defense, Deterrence, and the Prospects for Peace* in R. Hardin *et al., Nuclear Deterrence, Ethics and Strategy* (University of Chicago Press, Chicago, 1985) p. 287.
3. Michael McGuire, 'The Ultimate Umbrella' *Times Literary Review* (31 October 1986) p. 215.
4. S. Croft and P. Williams, 'SDI and the path from Reykjavik', *ADIU Report*, Vol. 8 No. 6, (November-December 1986) p. 3.
5. P. Garrity, 'The United States: the politics of strategic defence' *World Today*, Vol. 41, No. 1 (January 1985) p. 6.
6. *Time Magazine* (23 June 1986) p. 8.
7. North Atlantic Assembly, 'Interim Report', *Special Committee on Nuclear Strategy and Arms Control*, AD194 CS/AN (86) 2 (November 1986) p. 19.
8. *Common Security, A Programme for Disarmament* (Pan, London, 1982) p. 5.
9. *See* 'Reagan's Star Wars', *The New York Review* (26 April 1984) *passim*.
10. Lord Zuckerman, 'The Wonders of Star Wars', *The New York Review* (30 January 1986) p. 38.
11. For elaboration *see* I. H. Daalder and L. P. Whittaker, *SDI's Implications for Europe: Strategy, Politics and Technology* in S. J. Flanagan and F. O. Hampson, *Securing Europe's Future* (Croom Helm, London, 1986) pp. 56–57.
12. J. C. Toomay, 'The Case for Ballistic Missile Defense' *Daedalus*, **114**, No. 3 (Summer 1985) p. 236.

13. S. Croft and P. Williams *op. cit.* p. 1.
14. K. B. Payne, *Strategic Defense: 'Star Wars' in Perspective* (Hamilton Press, New York 1986) p. 96.
15. *See* M. Midgley, *Deterrence, Provocation and the Martian Temperament* in N. Blake and K. Pole, *Dangers of Deterrence* (Routledge and Kegan Paul, London, 1983) *passim*.

Part I
The Technological Basis

1 The Development of Command, Control, Communications and Intelligence Systems*

Group Captain Marten van der Veen

Man's continuing exploitation of space has led to the development of military tools of enormous value. And inasmuch as the 'national technical means of verification'—essentially reconnaissance and surveillance satellites—make it difficult for East or West to spring surprises on the other, these tools can contribute greatly to stability in the military balance, and hence to world peace. Huge strides have been made in space technology since the launch of the Soviet Sputnik on 4 October 1957.[1] In this first chapter we trace the story of these developments in the field of C^3I, and we outline what is technically feasible now, and what may prove possible in the near future.

Without doubt, one of the factors that has been crucial to our current capabilities in space-based C^3I has been the advance in rocket technology. The huge jump in satellite size from Sputnik's 83 kilograms to more recent satellites weighing 10 to 15 tons would not otherwise have been possible. But the advances in rocket technology have been well documented elsewhere, and will not be pursued further here.

We have also seen steady progress in the technology of electric power sources for satellites (essentially solar panels and nuclear generators), as well as in propellant systems for satellite manoeuvring. Such systems have become increasingly more reliable and—for an equivalent capability—both lighter and cheaper. But, almost by definition, it has been the strides in

* The views expressed in this chapter are those of the author and do not represent the position of the RAF.

electronics that have provided the key to current capabilities, and which have made it possible to put into effect the advances in communications theory foreshadowed by Shannon and others. It is these advances, and the capabilities they make possible, that concern us in this chapter.

THE TECHNOLOGY

Many of the wartime advances in radar formed the basis for postwar microwave technology. Such developments were key steps in the story of satellite communication systems. Microwave technology permits highly directional antennas, with beam widths only a few degrees wide, that are small in size and light in weight. And the high microwave frequencies allow us to use large band widths, so making possible communications at high data rates. The old wartime magnetrons which first powered centimetric radars led eventually to the travelling-wave tube (TWT) amplifiers of today—a valve much used in satellite systems.

The other advance in the field of electronics that has been so crucial to satellite systems was the development in 1948 of the transistor. The transistor could do most of the jobs of the thermionic valve, and it was cheaper, smaller and consumed far less power. The development of the transistor led within ten years to the integrated circuit, that is a complete circuit—rather than simply a discrete component—on a single wafer of silicon. The integrated circuit has revolutionised electronics in general, and communications and computers in particular. The story of the microcomputer and its development from the old first-generation all-valve computers is so well known that it does not bear retelling. But modern satellite technology would not be possible without the microprocessor in its many forms—all based on integrated circuits, implanted—as often as not—on the silicon chip.

Solid state technology has also had an impact beyond communications and computing. Solid state devices sensitive to visible or infrared radiation allow visual images to be converted into an electrical analogue—which can then be transmitted over a communications link. Scanning techniques can be used—by mirror or prism—to let a narrow strip of the image fall in turn on an array of photodiodes. But nowadays microelectronic silicon chips—each containing a number of light-sensitive elements—can be formed into larger arrays on to which the whole image can fall. These so-called charge-coupled devices (CCD) can be tapped off in sequence, again to provide an electrical analogue of the visual image.

However, the production of reconnaissance imagery is far from totally dependent on solid state devices. The early reconnaissance satellites used to eject canisters of film, complete with a radio homing beacon, for retrieval on earth. But subsequently, photographic imagery was processed in space, and converted by some form of scanning technique into video tapes whose content would be down-linked to earth at a suitable time over a communications link. As television cameras improved, their derivatives too—the vidicon and the return beam vidicon (RBV)—were used in space, again primarily with a view to providing video tapes for radio transmission back to earth.

These, then, are the main developments in the field of electronics that have made possible the vast strides in satellite capability for C^3I. We now turn to the various roles.

THE ROLES OF SPACE-BASED C^3I

The military C^3I roles are usually divided—somewhat arbitrarily—into:

- communications
- surveillance and reconnaissance
- meteorology
- navigation
- geodesy

However, as space-based systems become ever more integrated, the division between them will become increasingly blurred. We start with communications satellites, as the technology used is so vital to all the space-based C^3I roles.

Communications satellites

The advent of satellites provided a means of reducing our reliance on high frequency (HF) communications for long-haul communications. Useful as HF communications are (and certainly will continue to be), they suffer considerably from fading and interference; furthermore, with so many users in the comparatively narrow radio band between 2 MHz and 30 MHz, band widths have to be small and data rates are therefore very low. Although tropospheric scatter communications systems, as well as telephone landlines, can also provide beyond-line-of-sight communications, they too suffer from constraints. Satellites now offer worldwide communications links which do not suffer from fading or interference, are reliable at any time of day or night, are cheap in relation to their capabilities, and can operate between mobile terminals. During the early years, there was some debate about the merits of passive satellites that did no more than reflect the incident radio signal. Now active transponder satellites—which receive, amplify, and retransmit the incident radio waves—have become so reliable that they have totally supplanted passive satellites.

The first active-repeater communications satellite—Courier 1B—was launched in 1960. The first commercial communications satellite—Telstar—could receive and retransmit information simultaneously, and had a band width large enough to relay live television pictures. Subsequently, further active satellites were launched with improved technology—particularly in terms of lifetime, power and band width of the travelling-wave tube amplifiers. Nevertheless, satellite power levels were low, and because these early satellites were in low earth orbits—and so moved across the sky—they required large and complex tracking antennas on earth. Furthermore, such satellites were only visible to any particular earth station for a relatively short time during the day, so necessitating several satellites in order to maintain continuous communications. Fortunately advances in rocket technology soon permitted satellites to be placed in geostationary orbits.[2] As such satellites orbit the earth at the same rate that the earth is spinning, they remain stationary relative to points on earth. Not only does this simplify antenna pointing, both in space and on the ground, but it does much to dispense with the need for Doppler or power transmission compensation in the communications links. The first geostationary satellites, Syncom 1 and 2, were launched in 1963. The first such military communications satellite was Skynet 1a, launched in 1969—a great British achievement.

The self-evident advantages of satellite communications prompted a search for greater capacity. Although rocket-booster payloads were clearly going to increase, cost considerations drove the scientists and engineers to look for ways of increasing transmission data rates. By 1962 the Bell Telephone Laboratory had developed a commercial-scale system of pulse code modulation (PCM) based on solid state technology which provided the basis for more advanced digital encoding such as delta modulation. But the desire to permit many geographically dispersed terminals to use satellites to communicate with other dispersed terminals required radical new ways of accessing satellites. In the early days, a variety of terminals were routed through a primary earth terminal, but this soon became inadequate. Consequently multiple-access techniques were developed, allowing a number of ground terminals to access the satellite simultaneously without being routed through a primary earth terminal. Four methods are now in use: frequency division multiple access (FDMA), time division multiple access (TDMA), code division multiple access (CDMA) and space division multiple access (SDMA).

Using FDMA, each ground terminal is allocated a fixed frequency and a predetermined band width and power budget. Each receiving station must demodulate and demultiplex the channels of interest.

The concept of TDMA is based on assigning each ground terminal a time slot for transmission and reception. The time slots are short but frequent—to the extent that the effect of the process is imperceptible on a voice channel. TDMA makes for high transponder efficiency—as only one ground terminal has access at any one time, the whole band width of the transponder can be used. As might be surmised, TDMA requires the use of high speed digital processing equipment in space, and this has been possible only since the development of microprocessors in the early 1970s. The Canadian Satellite Communications System used a limited TDMA facility for the first time in 1975, and by 1980 Intelsat V had a full TDMA capability.

The successful deployment of microprocessors in space has led to even more complex signal-processing techniques which are particularly useful in the military field. CDMA has the triple advantage firstly of permitting multiple access by different terminals, secondly of providing good electronic counter-measures (ECM)-resistance to jamming as well as security against spoofing or exploitation, and thirdly of having an inherently low probability of intercept—and therefore being difficult to detect and jam. CDMA relies on complex signal processing and, because the resulting data stream is at a very high data rate, it requires a large band width for transmission. However, the resulting signal, together with those from other ground terminals, may be amplified and retransmitted by satellite transponders, and only a terminal with knowledge of one or other of the unique data streams is then able to reconstitute the original voice or data signal. In order to make the CDMA transmissions even less vulnerable to jamming, each unique data stream may also be used to alter the transmission frequency, or even the timing of microsecond transmission bursts.

SDMA relies on different ground terminals using different satellite antennas—implying that the ground terminals must be far enough apart to be in the footprint of only one antenna each. For military use, advanccs in antenna design have been crucial. The first operational military communication satellites—the Defense Satellite Communications System (DSCS) satellites first launched in 1966—were relatively simple spin-stabilised spacecraft, having few moving parts and designed for the highest reliability. The United Kingdom's Skynet 2 satellite in 1969 used a mechanically despun antenna with 17° beam width to provide earth cover. Since then antenna designs have progressed significantly. One of the largest communications antennas in space was the nine-metre parabolic reflector of ATS-6. Research is currently being carried out to devise lighter, more efficient, reflectors and it is now feasible to deploy thirty to forty metre antenna structures in space. But in recent years there also has been a trend towards higher frequency. Higher frequencies provide wider channel band widths, and permit smaller antennas with the high gain required to provide narrow beam widths. Rigid reflector technology is now being directed towards the development of techniques for producing accurate reflector surfaces, using new materials which are lightweight and able to withstand the high thermal stresses of space. Accuracy of reflector surfaces is critical in order to avoid out-of-phase relationships in the beam, which can lead to excessive side lobes which waste power and provide an entry for jamming signals.

All these advances have made satellite communications much more suitable for military use in a hostile environment, and largely explain their ever-increasing use.[3] Indeed, a staggering 70 per cent of all long-haul American military communications is said to be transmitted by satellite. The three main American satellite communications systems use satellites in geosynchronous orbits—the Defense Satellite Communications System (DSCS, already mentioned), the Fleet Satellite Communications System (FLTSATCOM), and the Air Force Satellite Communications System (AFSATCOM)—many of whose transponders are in fact carried aboard FLTSATCOM satellites. DSCS provides the high capacity secure voice and data links for the US World Wide Military Command and Control System (WWMCCS) linking the President, his Chiefs of Staff and his major commanders worldwide. The FLTSATCOM system, as its name implies, links mobile forces—particularly ships and some aicraft. The AFSATCOM system provides low data rate links—with correspondingly enhanced antijam performance—to control nuclear forces. Each of these three systems is apparently to be replaced or complemented by the Department of Defense (DOD) Milstar satellite system in the 1990s—satellites that will work at extremely high frequency (EHF).

Without the vast strides in electronics none of these advances would have been possible. The revolutionary steps in C^3I have been taken. Future developments are likely to be evolutionary, rather than revolutionary, as attempts are made to improve satellite life and performance.

Solid state technology continues to improve, in providing higher power low noise amplifiers and oscillators working at ever higher frequencies. Gallium–arsenide (GaAs) field effect transistors (FETS) are the most advanced in development, but they are likely to be limited to 40 GHz; above this impact avalanche transit time (IMPATT) diodes will probably be used. They are likely to serve as the building blocks for transponders in the 40 to 70 GHz region. Impressive progress is also being made in the development of monolithic microwave integrated circuits (MMIC) since the first GaAs MMIC amplifier built in 1975. The degree of integration is likely to increase much further in the 1990s, so providing evermore compact mixers and oscillators working at ever higher frequencies. The classic TWT amplifiers are also improving in terms of power output, frequency range, efficiency and life expectancy. Progress in electron beam amplifiers and fast-wave devices could well extend the frequency range into hundreds of GHz. Such progress will increase communications band width, and hence capacity and ECM resistance.

In parallel, advances are certain to continue in antenna design. Modern multibeam antennas are highly versatile, both in beam-shaping control and in frequency usage. Furthermore considerable interference rejection can be achieved. New materials are being examined for the construction of microwave lenses to provide wide band widths, and considerable work continues to be carried out on phased-array antenna systems. As they can provide electronically steerable beams they are very suited to satellite systems. However, phased-array systems are currently rather heavy and costly, and—on account of their complexity—require quite high power signal-processing control systems. In the same way, microstrip antenna technology is likely to become more important in the future. MMIC devices appear to have considerable potential in this area.

Future advances in these three main areas—solid state devices, TWT design, and antenna design—promise to provide even more capable satellite communications systems. And the Strategic Defense Initiative will undoubtedly boost such progress. But in view of their great importance in the military domain, it is clear that both East and West will do all they can to reduce the vulnerability of their satellites to ECM or to physical attack.[4] Indeed modern military satellite design features hardening against laser weapons and against electromagnetic pulse to enhance survivability during a crisis or in war. Nevertheless resistance to jamming—as has been outlined—is perhaps more readily achieved than hardening against physical attack. Magnificent as the strides have been in providing such reliable and instantaneous satellite communications over the past twenty years, most analysts therefore believe the remaining potential vulnerabilities of satellites to antisatellite weapon systems will require the retention of many of the older HF, tropospheric scatter and landline systems to provide a degree of redundancy.

Surveillance and Reconnaissance

From time immemorial military commanders have wanted to know what was on the other side of the hill. In the nuclear age the requirement for intelligence has become even more important. For a defensive alliance like NATO—heavily reliant on reinforcements—early warning of Soviet military moves is essential.[5] And for the longer-term maintenance of a military balance—both conventional and nuclear—satellite reconnaissance can provide vital intelligence about Soviet weapon systems and deployments—intelligence that would, given the Soviets' closed society, be impossible to obtain any other way. Indeed many of the US/USSR arms control agreements implicitly or explicitly rely on these 'national technical means of verification'. These factors have prompted the Americans to develop numerous surveillance and reconnaissance satellites for a variety of tasks. The Soviets have similar programmes.[6] Again subdivision is worthwhile into satellites that can provide:

images (by photographic, radar or infrared means)
electronic intelligence (ELINT)
indication of nuclear detonations
early warning of Intercontinental ballistic missile (ICBM) and SLBM launching

High quality photographic reconnaissance by satellite developed rapidly in the 1960s as the ever-increasing booster payloads allowed both sides to launch large satellites with the long focal length cameras required for high resolution. In those early days, the quality obtainable from photographic film was much better than that produced by television cameras. Consequently the majority of the so-called 'close-look' satellites periodically ejected film capsules for recovery on earth. It is clear from the satellite imagery that has been released to the open press that a resolution of less than one metre was quite rapidly attained from satellites at, say, 180 km altitude. Reconnaissance satellites were often put into highly eliptical orbits, with the perigee—the lowest point in the trajectory—over the reconnaissance target; such an orbit would minimise the time at low level and hence the atmospheric drag on the satellite. Whilst the resolution obtained would not quite permit *Pravda*'s headlines to be read from space, as some have suggested, it certainly became possible to distinguish military equipment. Verification of arms control agreements became possible, as did the ability to monitor military operations in crisis areas—as both sides appear to have done during the Middle East Wars and the Falklands War for example.

The problems of photographic film recovery by capsule, and the inevitable delay in doing so, prompted the search for means of transmitting the photographic information by radio. Systems were produced to develop the photographic images in space before passing the film through a scanning system (similar to a television camera) which converted the photographic image into an electrical analogue, albeit with some degradation in quality. The electrical data stream could then be transmitted to earth, normally after some form of storage on magnetic tape until the satellite came within line of sight of a friendly ground terminal—although more recently the use of radio-relay satellites can overcome this delay. Satellite photography was, of course, not confined simply to the visible spectrum. Photography in the infrared bands is also much used for civil purposes. Indeed, 'false colour' images can be obtained from the various infrared bands to indicate the relative distribution of reflected radiation—and this has proved a very useful mechanism in mapping earth resources by satellite, as Landsat has done.

However, the advances in electro-optical technology in the 1970s prompted the use of such devices in space. The development of the RBV for high quality TV pictures, and devices such as multi spectral scanners (MSS), began to reduce the dependence on photographic film for high resolution images. And as both RBVs and MSSs permitted image transmission digitally, detailed near real time reconnaissance became possible. RBVs are, apparently, used for 'area surveillance' rather than for 'close-look' reconnaissance. Recent years have also seen the introduction of CCD sensors which offer the capability of high resolution, high sensivity and high reliability with minimal size, weight and power consumption. Essentially these are microelectronic silicon chips containing a number of light-sensitive elements which can be arranged in a CCD array. The resolution capabilities of these devices seems to be limited only by the technology of the optics and the detector size (which can be tailored to the need)—although the data rate of the communications channel may continue to be a limiting factor.

Advances in thermal-imaging techniques have also proved remarkably useful for space-based systems—again primarily in civil satellites. Rather than simply relying on reflected infrared energy, thermal imaging relies on the longer infrared wavelengths. The temperature of the infrared detectors is a crucial factor which sets limits on the ability to detect faint infrared sources. These arrays are often termed 'staring' focal plane arrays, and are clearly able to build up a complete image more quickly than can thermal imagers working on the line-scan principle.

The last decade has seen major advances being made in image enhancement techniques. Photographic techniques have long been used to improve contrast, but a major step forward has been the introduction of computer enhancement of the images in digital form. Computer techniques can now also increase contrast by 'edge enhancement', or by artificially stretching the so-called 'grey scale' so making imagery of a dull, misty scene as sharp and clear as if it were taken on a bright, clear day. Other computer techniques compare changes in the digital image between one pass of the satellite and the next. Such mechanisms can greatly enhance the resolution of the raw data.

But unfortunately such systems all rely on visible light or frequencies close by. The use of the shorter infrared wavelengths does provide some capability by night, but cloud cover—all but the lightest—renders even such systems unusable. Both sides are therefore turning to centimetric and millimetric radars. However, the use of radar implies increased power requirements; and frequencies so much lower than visible light imply large antennas to ensure reasonable resolution. Some Soviet radar satellites consequently appear to use nuclear power sources. Conventional radars, unfortunately, are limited in resolution by the beam width of the antenna. Nevertheless, some side-looking radars, using long slender antennas to achieve the narrowest beam width in azimuth, have been put into space. But the concept of synthetic aperture radar (SAR)—whereby the movement of the radar transmitter through space effectively substitutes for a large antenna—is probably the key to the further development of radar satellites. The correlation of radar returns from an object on earth as the satellite moves in its orbit does, however, require considerable computing power. As the devices to provide this have been developed only quite recently, it should not surprise us that SAR satellites have become practicable only in the last decade, even though the principle of SAR has been well known for many years. In America the first SAR radar development programme was initiated in 1953 under the title 'Project Wolverine'. Analogue correlation techniques for SAR were proved in the 1960s, but it was not until the 1970s that real time digital SAR was demonstrated for the first time. The SAR images that have been released (from the American Seasat, for example) are impressive in both clarity and resolution—and are similar in quality to high altitude photography. Radar satellites can be used for ocean surveillance. The introduction of long-range antiship cruise missiles on both sides has prompted a search for a mechanism to detect surface-vessel launch platforms at long range. Radar satellites can meet this need, and can of course be directed to look at likely targets by other satellites incorporating infrared and ELINT sensors. The detection and tracking of aircraft could also provide commanders with vital early-warning information; the Americans appear to be devoting considerable efforts to this, as their HI-CAMP and Teal Ruby programmes indicate.

One of the difficulties with high resolution 'close-look' satellites is knowing what to look at. Some satellites have been launched for area surveillance, with optics providing very wide swathes of reconnaissance coverage. Indeed, some reconnaissance satellites now do both close-look and area surveillance. But both sides have also long deployed ELINT satellites, which can in part also meet this need. Such satellites contain receivers for detecting the adversary's radio and radar transmissions. Triangulation can be used to locate the source, and so cue imaging satellites when required.

But ELINT satellites have other uses too. They can—even without the back-up of imaging satellites—give a good indication of the deployment and readiness state of forces, both land, air and maritime. They can provide essential information on the electronic characteristics of enemy systems to permit the design and adjustment of antiradiation missile homing heads or of ECM systems to jam them. And, indeed, information on the disposition of enemy systems would be crucial for the routing of friendly aircraft or Cruise missiles. Finally, it is clear that such satellites are vital for the collection of telemetry data from ballistic and other missile tests.

Both sides have also deployed surveillance satellites to detect nuclear detonations. Originally these were designed to ensure the other side's compliance with the Partial Test Ban Treaty, which prohibits tests in the atmosphere. Such satellites have X-ray and gamma ray detectors, and possibly some form of spectrometer able to detect the presence of typical fission products. Some are said to incorporate TV cameras which can provide imagery to guard against false alarms. Whole earth coverage is clearly essential for such a system—which therefore has to rely on a number of satellites. Such systems are apparently now also considered essential for reporting wartime detonations over one's own territory (to provide the facts needed to determine the scale of response), as well as over one's opponent's territory (to provide feedback on the results of one's response).

The last category of surveillance and reconnaissance satellites comprises early-warning satellites—early warning, that is, of enemy ICBM and SLBM launches. The satellite sensors detect the enormous amount of infrared and visible radiation produced by the rocket motor plume during the boost phase. The first attempts at producing such a capability in the late 1960s were apparently not fully successful. However, improvements in sensor reliability and sensitivity have permitted the Americans (and presumably the Soviets) to deploy a fully functioning system providing 30 minutes warning—approximately double that provided by terrestrial systems like the ballistic missile early-warning system (BMEWS).[7]

Meteorology

Satellite systems have without doubt revolutionised weather forecasting, and they have become important in detecting the limits of pack ice, icebergs and new ice formation. This in itself has been a most useful advance for the planning of military operations. But equally important is the fact that satellites permit the surveillance of the weather over the adversary's territory so permitting wind-compensating corrections to ballistic missile trajectories in war.

Navigation

The need to know one's exact position has long been a feature of military operations, particularly at sea and in the air. Whilst huge strides have been made with terrestrial systems, some lose accuracy as time passes (inertial navigation systems), and others provide only regional cover (LORAN, Decca and Tacan). Only Omega, recently introduced, provides a reasonably accurate worldwide fix. But it is a system dependent on eight radio sites in seven independent countries, and it could not therefore necessarily be relied upon in war. Because their trajectories are precisely known, satellites have long offered the potential for an accurate navigation system for use by terrestrial vehicles. In essence, one can fix one's position on the earth's surface if one knows one's distance to each of three satellites (from the length of time taken for a radio signal to arrive) which are continuously broadcasting details of their own positions. Such a system permits a very quick determination of one's position; an alternative, but slower method, is to rely on measuring the change in Doppler shift in a satellite signal over a period of a number of minutes.

The United States first deployed the Transit satellite navigation system in 1964 to provide the US Polaris submarines with a means of periodically updating their on-board initial navigation systems. For Polaris missile trajectories to be correct, it is essential that their launch positions are known with considerable accuracy. The Transit system (which will be phased out in the 1990s) was released for civilian use also in the late 1960s. It now provides every user—both military and civilian—with a navigational accuracy of 50 to 500 metres anywhere in the world—depending on the type of receiver and the user's knowledge of his own velocity. The replacement system—the United States Navigation and Satellite Timing and Ranging Global Positioning System (NAVSTAR GPS) is intended to provide a horizontal accuracy of ten metres circular error probable, and a three-dimensional accuracy of sixteen metres spherical error probable, when all eighteen satellites equally spaced in six orbital planes are in service in 1989 or so. A twin-signal structure will be used, providing coarse navigation for civilian users, and precise navigation for military users who have the correct access code. According to some sources the navigation will be sufficiently precise to enable aircraft to release their weapons blind. NAVSTAR GPS will also provide a key service to naval surface vessels and to army units, as well as enabling other satellites to determine more accurately their own position—and so also that of the objects detected or surveyed (so improving weapons targeting). It will probably also be used to correct ballistic and Cruise missile navigation systems.

Geodesy

The earth is, of course, not a perfect sphere. The earth's gravitational field is therefore by no means constant, to the extent that the trajectories of ICBMs and SLBMs can be perturbed by the variations. Both sides, therefore, have an interest in determining the exact shape of the earth—and indeed in improving the accuracy of their maps for missile-targeting purposes amongst other things. Geodetic satellites can use a variety of sensors, including radar altimeters, to provide information on the precise size and shape of the earth. In this way they make a vital contribution to military preparedness.

CONCLUSION

In so many areas satellite C^3I systems now enable us to do what was not possible before. Communications satellites provide both strategic communications from headquarters to headquarters, as well as long-range tactical communications to individual units, ships or aircraft. Peacetime satellite reconnaissance permits us to keep a watch on the other side, and so maintain a stable military balance. And in the future satellite systems are likely to provide ever increasing factual reconnaissance in real or near real time. Early-warning systems vastly reduce the chance of a surprise attack, and so also contribute to world peace. Satellite navigation and communication systems also confer enormous benefits on the civilian scene. None of these tremendous advances would have been possible without the vast strides in electronics, and particularly the development of the silicon chip in all its forms.

NOTES

1. A good review of satellite development up until the late 1970s can be found in Bhupendra Jasani, *Outer Space-Battlefield of the Future?* (Taylor and Francis, London, 1978).
2. The number of 'slots' available for geostationary orbit are now very few. *See* Joseph N. Peltan, 'Gold rush in the Clarke orbit', *Society* 21 (January/February 1984) pp. 8–12.
3. For a current assessment *see* Norman West, 'Military satcoms—risks and implications', *NATO's Sixteen Nations* 29 (November 1984) pp. 34–38.
4. *See* Office of Technology Assessment, *Anti Satellite Weapons, Countermeasures and Arms Control*, OTA-ISC-281 (GPO, Washington DC, September 1985).
5. *See* the entire volume of *NATO's Sixteen Nations, op. cit.* for an assessment of the Alliance's dependence on space assets.
6. *See Soviet Space Programs: 1976–80* (US Senate, Committee on Commerce, Science and Transportation, May 1985).
7. *See* Bruce Gumble, 'Air Force upgrading defences at NORAD', *Defence Electronics* (August 1985) pp. 85–108 for a review of the extensive modernisation of the US early warning system including BMEWS.

SOURCES

Space Handbook (AU-18), Lt. Col. Curtis D. Cochran; Lt. Col. Dennis M. Gorman; Maj. Joseph D. Dumoulin, Maxwell Air Force Base, Alabama 36112–5532, January 1985.

Countdown to Space War, Bhupendra Jasani and Christopher Lee, Taylor and Francis, London, 1984.

Outer Space—A New Dimension of the Arms Race, Bhupendra Jasani (SIPRI), Taylor and Francis, London, 1982.

Outer Space—Battlefield of the Future, Bhupendra Jasani (SIPRI), Taylor and Francis, London, 1978.

The International Counter-measures Handbook (11th Edition), Charles H. Wiseman Editor-in-Chief), EW Communications Inc., New York, 1986.

US Space System Survivability, Robert B. Giffen, The National Defense University, Washington D.C., 1982.

2 Strategic Command and Control and the Militarisation of Space

Michael Clarke

Command, control and communication (C^3) for strategic forces has always been more art than science, and fairly abstract art at that.[1] While both superpowers have concentrated heavily on the development of strategic weapon systems they have neglected the C^3 which is necessary to back them up. It is not difficult to understand why this has happened; in a world of scarce resources attention naturally focuses on the weapons. It is, however, difficult to understand why it is that this situation persists when C^3 deficiencies have been clearly exposed as glaring, comical and dangerous;[2] dangerous not just for the superpowers but also for the rest of us who depend for our survival on a high degree of nuclear crisis management on the part of Washington and Moscow. It is even more difficult to understand why it is that the United States, in particular, seems about to repeat—yet again and this time on a scale hitherto undreamt of—exactly the same mistake of developing the weapons systems above the level of the necessary C^3. Indeed, it is not too much to say that the militarisation of space is being driven by the prospects of weapons technology but the gains and losses, both for the superpowers and for the rest of us, will revolve around the effects which this militarisation has on strategic C^3. This will be what space competition is really all about in twenty or thirty years.

A final puzzle is that the US is pushing ahead with this development of space technology when by almost every realistic scenario it must lose to the Soviet Union in C^3 terms. *Any* militarisation of space will make life harder for American strategic planners. There are two possible reactions to this fact. One is to attempt co-operation with the Soviet Union in the interests of crisis management and strategic stability and pursue a vigorous policy of arms control that would be designed to safeguard clear and reliable C^3 facilities in space. This would require much more dramatic superpower agreements than were achieved twenty years ago in the golden age of arms control. So far, there is no sign of this becoming a reality, and since 1983 very little evidence of an American willingness to move in that direction. The other reaction is embodied in the SDI, which attempts to capitalise upon the United States's present lead in space, and space-based C^3, in order to outrun the Soviet Union in this technology.[3] The attempt to do this seems to be doomed to failure; indeed, from the point of view of C^3, the SDI has all the makings of another 'MIRV debate' where the US stimulated multiple independently targetable re-entry vehicle (MIRV) development to capitalise upon its technical superiority, only to become the loser as all ground-based targets eventually became vulnerable to less sophisticated, but very effective, Soviet MIRVs.[4]

The SDI programme places US C^3 in a situation from which it cannot gain in the long term, and which is likely to make C^3 a force for destabilisation rather than stabilisation in the strategic nuclear relationship. To substantiate these assertions, this chapter will examine existing US and Soviet C^3 and then turn to the implications of future developments: in particular the growing dependence on space-based C^3, the implications of SDI on ASAT, and the implications of SDI for future requirements of strategic battle management.

STRATEGIC C^3 IN THE MID 1980s

Neither of the superpowers operates what could be characterised as a strategic C^3 'system'. Both of them have an enormous complex of arrangements covering and overlapping many functions with different applications of technology. In general, however, we may conceive of C^3 as a matrix between technologies and functions. The technologies cover ground-based facilities, airborne facilities and space-based assets. All have their own strengths and weaknesses and feed into each other in a mixture of ways: satellites have to have their ground stations to receive messages, as do aircraft; and aircraft have to duplicate ground-based command facilities in order to offer some survival against the likely destruction of all ground-based facilities in a nuclear exchange. Different branches of the armed services place different patterns of reliance on these facilities. The US Navy, for instance, makes more extensive use than the army of satellites for its ship-to-ship and ship-to-shore communications. Soviet forces make more use of secure landlines than does the US. Generally speaking, all of these technologies are used in one way or another to perform all the major C^3 functions.

The 'functions' side of the matrix can be characterised under four headings. Firstly, there are the peacetime functions, which generally consist of reconnaissance and monitoring, communication, navigation, meteorology and geodesy.[5] Inevitably these are the functions which receive the most immediate attention and finance and which are being more constantly updated; they are the only functions which are consistently in operation. In some respects the performance of these functions militates against the performance of wartime functions, since secure and resilient systems are not normally the easiest to operate. There is a tendency to invest more heavily in relatively soft 'peacetime' C^3 technology in the hope that it can be adapted to wartime requirements if necessary. The second category of functions covers early warning of strategic attack. This involves detection of missile launches, tracking, invocation of national command authority (NCA) resulting in a top-level political decision, target estimation and—often neglected as a genuine problem—confirmation of nuclear detonations as they occur. Thirdly, it is necessary to be capable of launching a retaliatory strike. This requires survivable communications networks—an immense task—and launch procedures which can be technically autonomous, so as to continue to function amid a great deal of damage. Finally, there is a battle-management function, which requires a continuity of communication and political decision making, the control of existing forces and the accurate monitoring of damage to both the homeland and the C^3 system itself. To say that nuclear battle management is a function of C^3 is not to imply that it can necessarily be performed. There is widespread scepticism as to whether *any* effective battle management C^3 would survive a nuclear attack. Nevertheless, battle management is an aspiration which grows out of a long-standing American urge to consider flexible response and selective nuclear options as a possibility.[6]

All of these functions employ space technologies in one way or another. The US presently runs eight systems of military communications involving at any one time more than twenty operational satellites, most of which are in high geosynchronous orbits at 22,250 miles.[7] The most important of these are the DSCS II and DSCS III, FLTSATCOM and AFSATCOM, though the AFSATCOM system is accommodated on other satellites.[8] There are four navigation networks which at present deploy more than twelve satellites, though this will expand to at least twenty-seven by 1988.[9] Four satellites are used for meteorology and six for ocean reconnaissance. Early warning is encompassed in the DSP (Defense Support Program) which consists of three geosynchronous satellites containing an infrared detecting telescope. Electronic intelligence listening is carried out by four Rhyolite satellites whose primary job is to intercept the telemetry from Soviet missile tests.[10] Sensors to detect nuclear explosions are carried on two Vela Hotel satellites, though they are also on DSP and SDS (Satellite Data System) satellites and will be overtaken by advanced detection systems on NAVSTAR, the new navigational system. Finally, photographic intelligence and geodesy are covered by a variety of satellites about which it is hard to obtain unclassified information. Certainly there is a KH-9 (big bird) and KH-11 satellite and there may be at least four of these in orbit at any one time.[11] There are two geodesy systems, GEOS-3 and LAGEOS, but it is not known how many satellites are deployed for them.

Soviet satellites cover most of the same C^3 functions. There are three communication systems requiring around fifty satellites in various orbits.[12] There are fifteen or more satellites at any one time supporting two navigation systems, one of which—GLONASS (Global Navigation System)—will be an equivalent to NAVSTAR; four ocean reconnaissance satellites and four meteorological satellites; nine early-warning and six electronic intelligence satellites. Again, photographic reconnaissance and geodesy are difficult to pinpoint, but the Soviet Union launches over fifty short-lived reconnaissance satellites in most years and a (classified) number of geodetic ones.

The amount of hardware in orbit is, of course, no guide to performance. As far as that is concerned, the United States is way ahead of the Soviet Union, though still way behind an adequate level of strategic C^3 to match its weaponry. It is ahead of the Soviet Union on almost every technological score. US satellites remain operational for much longer than Soviet ones; up to ten years in many cases, whereas Soviet satellite life can be measured in weeks or months. American satellites are more reliable and sophisticated. The Soviet Union has had to use existing designs in modified forms; it has a very poor record in microengineering resulting, for instance, in poor guidance and sensor technologies, and frequent and debilitating malfunctions.[13] Most important, Soviet satellites are predominantly in low or very elliptical orbits[14] which make them less useful in operational terms and certainly more vulnerable to low orbit antisatellite weapons which are now almost at hand. The Soviet Union has about 75 per cent of its satellites in low, vulnerable orbits below 12,000 miles. The United States has only 40 per cent at the same altitude and these are photoreconnaissance satellites, which *have* to operate in low orbits, and satellites performing peacetime functions such as navigation, ocean reconnaissance and meteorology. The most sensitive American C^3 functions—communications, electronic intelligence and early warning—are performed almost entirely at geosynchronous orbits, which will remain invulnerable for the immediate future. The United States, then, is in a commanding position in this respect.

The second part of the proposition, however, is equally true: US C^3 is extremely deficient in relation to the tasks it needs to perform. The satellites themselves are inherently soft, with little protective shielding and sensors which are highly susceptible to interference and disruption. Nuclear explosions in space would create X-ray and gamma radiation that can destroy sensors, darken optical fibres and disrupt integrated circuits. Even radiation-hardened circuits remain susceptible to electromagnetic pulse. In the case of satellites it would take the form of a system-generated electromagnetic pulse (SGEMP), which occurs when a flow of current strikes the surface of a satellite, dislodging electrons which disappear into space, leaving behind a positively charged satellite.[15]

Another vulnerability, apart from the satellite itself, is its communication with the ground base. To date, tests indicate that interference in ground-to-space communication caused by nuclear explosions is relatively concentrated and short lived. A more serious hazard is deliberate jamming.[16] The essence of the problem, however, is the state of the ground-based stations. If *their* performance was impaired, as seems highly likely in a nuclear exchange, it is probable that satellites would lose their receiver of data, or their ground controllers on whom they depend for corrective action.

Though satellites in themselves are very vulnerable, they are less vulnerable at the moment than ground-based or airborne C^3 assets. Yet they are critically dependent on the survival of ground and airborne systems to make some use of their survival. Ground control stations, however, are wide open to direct attack with accurate warheads. Nor are there many of them to cover. The US navy's worldwide ground-based communication facilities add up to about fifty targets in all,[17] and there are fewer than 100 critical fixed C^3 targets in the United States,[18] perhaps 400 primary and secondary targets altogether.[19]

This is compounded by the fact that facilities are physically concentrated. Minot airforce base in North Dakota, for instance, houses three Minuteman III squadrons (150 missiles), their fifteen launch control centres, a B-52 strategic bomber wing, an F-16 fighter wing and a KC-135 tanker wing. Ellsworth airforce base includes a similar concentration in addition to one of the key airborne command posts.[20] Landline communications between ground centres would be severely disrupted. Most military networks are carried by the commercial telephone system.[21] Long-distance radio communication, particularly HF and VHF frequencies, are highly susceptible to the effects of nuclear detonations. As Blair put it so succinctly, 'As of 1985 Soviet strategic forces could quickly overwhelm virtually all ground-based C^3I'.[22]

Airborne networks, once the back-up to the ground-based C^3 system, are now the front line. The vulnerabilities of this, however, are staggering. The most significant airborne networks are the National Emergency Airborne Command Post (NEACP) to get the President and the Joint Chiefs of Staff safely into the air, the CINPAC ACP (Commander in Chief, Pacific), CINCLANT ACP (Commander in Chief, Atlantic), SAC ACP (Strategic Air Command), SAC relay aircraft, SAC airborne launch control aircraft, and TACAMO submarine communications aircraft.[23] Once airborne these aircraft would establish their own communication links with the ground, with each other and with satellites. However, it is quite likely that many of these aircraft would not get away before their bases were hit. Their escape times are of the order of fifteen minutes—way outside the eight minutes or so that SS-N-6 warheads would take to reach the bases from Soviet submarines that are known to patrol off the coast of Bermuda. Until 1983 the President's NEACP aircraft are based at Andrews airforce base outside Washington. It would take nine minutes for the aircraft to get into the air, and about the same time for the President to reach it via helicopter from the Whitehouse. Since the aircraft was clearly vulnerable, it is now based at Grissom, Indiana. The distance buys it extra time. Now the aircraft is safe but what about the President? Even having escaped an attack, the operations of these airborne command posts would be very questionable. Up to 1983 only two NEACP aircraft were hardened against the effects of EMP; all the others are vulnerable. The eighteen TACAMO aircraft have no facilities for in-flight refuelling, so their range and endurance are strictly limited.[24] All the aircraft will have to cope with fallout, heat and the atmospheric effects of nuclear explosions on the ground. Radio links will be highly problematical. To communicate with DSCS-3 satellites a super high frequency (SHF) must be used, but only the E-4 NEACP aircraft can accommodate the bulky equipment for SHF. Other aircraft in the airborne networks must use UHF equipment which links into the Lincoln Experimental Satellite

(LES), but UHF communication is easily blocked out by a high altitude nuclear explosion. UHF communications would go blank for several hours. In general, satellite communications, 'would make only marginal contributions to airborne force management'.[25]

Inevitably, less is known about Soviet C^3, though it is structured in a similar way to that of the US.[26] It is accepted that there is a massive network of civil defence, underground shelters, buried landlines and radio relays for ground-based communication.[27] The Soviet Union deploys nine early-warning satellites but appears to rely more on ground-based radar stations: the first layer consists of two over-the-horizon radars, the second of a network of eleven 'Hen House' radars at six locations on the periphery of the Soviet Union. A new network of six phased-array radars will provide all-round coverage for the Soviet Union by the 1990s.[28] Such ground-based assets, however, are just as likely to suffer from the same problems as beset the United States. There is a small fleet of airborne command posts (and some shipboard command posts) as a back-up, but since this is known to be rather more primitive than the US network, its problems and reliability can only be guessed at.[29] Generally speaking, the Soviet Union deploys more ground-based, more centralised, less sophisticated, more redundant and, in certain limited respects, more survivable C^3 than the United States.

THE GROWING DEPENDENCE ON SPACE

It is clear from the preceding sketch of strategic C^3 that the whole business has become something of a nightmare. Missile accuracy, forward deployments, the known effects of nuclear explosions and the unknown effects of nuclear explosions in operational conditions have combined with decades of relative neglect of C^3 to produce arrangements which are barely comprehensible. Peacetime C^3 functions have clearly improved and this might be regarded as a stabilising development. Wartime C^3, however, hardly inspires confidence and this may well spill over into a declining confidence in crisis management.

Space-based C^3 plays a significant though not an overwhelming role in present arrangements. It is perhaps inevitable that as ground and airborne networks become more doubtful the use of space-based systems will increase as the least unattractive option for the future. In this respect the SDI has a certain superficial logic to it. Space is being militarised anyway. The SDI is merely an attempt to stimulate and make coherent a thrust of technology which is inevitable. SDI makes very considerable C^3 demands in precisely the areas which play to American strengths over the Soviet Union. What could be more prudent in an uncertain world than to capitalise upon one's strengths and achieve a decisive lead in the militarisation of space?[30]

This line of thinking also helps rationalise a number of other developments which have tended to shift C^3 assets towards space. In peacetime, satellites offer much more flexible and efficient communications so that 70 per cent of US long-haul traffic goes via space. Then again, the withdrawal from some military bases, plus the desire to maintain a global military capability, has increased the incentive to use satellites.[31] The workload of C^3 has increased enormously with weapon complexity at a time when new applications for space-based C^3 are becoming apparent. As Paul Stares points out, satellite C^3 is not in the front line in a conventional conflict in Europe, since there are many ground and airborne C^3 systems for European operations. But attrition rates would probably be very high in any European conflict so that satellites may rapidly become front-line C^3. And any new battlefield strategies which try to 'look deep' and 'strike deep' in Europe will perforce be more dependent on satellites for target acquisition and missile guidance.[32]

Above all, spaced-based C^3 seems to offer a way of mitigating some of the worst deficiencies of existing arrangements. The most important modernisation programmes involve increasing the survivability and hardening of all assets—ground, airborne and in space. Mobile ground control centres may well be introduced, and a great deal of faith is being put in a new generation of eight satellites known as MILSTAR.[33] The MILSTAR constellation will be deployed by the early 1990s and will provide the *primary* wartime communications link between aircraft, submarines, satellites and ground bases. It will operate on EHF, which will give it considerable hardening against EMP, SGEMP and jamming. It will reportedly have some manoeuvring capacity that will make it a more difficult ASAT target, and it will be able to function autonomously. It is even expected that strategic missile submarines will be able to use it and reply to, and acknowledge, orders received. This is an impossibility at present since any reply from a submarine would expose its position. The fact that submarine communications are predominantly one way poses a unique inflexibility on existing battle plans.[34]

MILSTAR will be a multipurpose satellite communication system. If it works as advertised, it will perform the majority of crucial communication functions and will take over all the AFSATCOM and some of the FLTSATCOM functions, and the Satellite Data System (SDS). It could also have some capability for flushing bombers and command aircraft sufficiently early. MILSTAR, a $10 billion programme, is the centerpiece of a C^3 rationalisation.[35] Programmes to provide some back-up alternatives should MILSTAR fail under operational conditions are costed at less than $30 million and are in reality no more than a series of general ideas.[36]

All of this seems eminently sensible from a US point of view. Its most important satellites are in safe, high orbits; they are being rationalised by MILSTAR, and the ground-based facilities are being steadily upgraded. Peacetime functions are performed by sophisticated satellites which provide the US with excellent intelligence under normal conditions. Indeed, President Johnson once noted that the intelligence information obtained from photosatellites alone had been worth more to the US than the cost of the whole space programme.[37] There are, however, two major objections to this eminently sensible reasoning.

Firstly, it is clear that the United States, more than the Soviet Union, is becoming dependent on space-based C^3 assets. Paul Stares has pointed out that statements about this 'must be interpreted with some care'.[38] In peacetime much satellite communication is not vital; the US chooses to use space because it is more convenient. It does have alternatives. This is undoubtedly true, but it is only true to an extent. US satellites tend to be multifunctional so that there is a considerable concentration of performance in a small number of units.[39] They also tend to be multipurpose, since they serve both strategic and conventional forces, and in many cases the civilian organisations which own the satellites and from whom the military lease some of their capacity. More importantly, if the investigations of Blair, Steinbrunner, Bracken, Ball, Arkin and Fieldhouse are at all accurate, then it is clear that in a strategic context, the US must regard its ground and airborne C^3 assets as tenuously survivable.[40] The choices of C^3 deployment which the US can exercise in peacetime may simply not exist in wartime.

The Soviet Union aims to use its satellites as a back-up for its ground and airborne systems. Soviet ground-based systems may be more resilient than American ones, and as a continental land power its task in maintaining them is simpler. For the US, this situation is reversed.[41] There is, in short, general agreement that the US is more dependent on space-based C^3 than the Soviet Union, certainly in peacetime, and almost certainly in wartime. And this trend is likely to continue and strengthen. Nuclear submarines, for instance, are becoming increasingly dependent on satellite communications for their safety.[42] Even Edward Teller, the architect of SDI, stated to the Senate Armed Services Committee in 1983; 'to talk about Star Wars is almost an inappropriate discussion of what the question is . . . to destroy space objects is relatively easy. We need space. We need space badly if for no other reason than for observation, knowledge when an attack is launched.'[43]

Secondly, does this degree of dependence matter? The answer is an unequivocal 'yes' if the militarisation of space proceeds on its present course. The Soviet Union appears to place great emphasis on its ability to strike at C^3 targets as part of its general operational art.[44] Certainly Soviet targeting policy in Europe and its deployment of Yankee class missile submarines close to the American coast indicates that Soviet planners have every intention of holding US C^3 at risk wherever possible. The next question is, can the Soviet Union hold at risk US satellites which will be hardened and secure in their high geosynchronous orbits? Can they do it if the US also derives the advantages of SDI technology for defence in space in the medium-term future? The answer is an almost certain 'yes', not in spite of SDI, but rather *because* of it. For whatever else it may or may not achieve, SDI will increase the performance of ASATs and from this the Americans can only lose. At best, the US could hope for a stand-off at parity, in which *all* space-based C^3 is vulnerable and a new deterrent relationship, based on new triggers, comes into being. Why is it that SDI cannot help but promote ASATs?

SDI AND ASAT

The close relationship between SDI and ASAT arises for both technological and political reasons. The technology of ASAT began to seem promising after 1977, certainly in relation to what had seemed possible previously.[45] Both superpowers have moved tentatively towards ASAT capabilities. The technology is very uncertain and the disincentives to doing it the easy way by installing nuclear weapons in space have been backed up by the Outer Space Treaty 1967 and the Anti-Ballistic Missile Treaty 1972. Nevertheless the United States has tested a first-phase ASAT, consisting of a two-stage booster rocket, carried to a high altitude aboard an F-15 fighter. The interceptor is a miniature homing vehicle which locks on to its target and collides with it. It was first tested in January 1984 and could be operational in 1988. The Soviet Union has tested a far less flexible ASAT which is launched from the ground and has to get itself into an adjacent orbit to its target so that it can explode near enough to kill it with shrapnel. The system is very limited, however, in the targets it can hit; it has no capacity to attack most US satellites in high orbits and takes a long time to get into operation. Although nine of twenty Soviet tests conducted since 1968 have been successful, its more modern optical/infrared homing device has proved to be a failure and has failed in at least five out of six tests since 1977.[46]

Neither of these systems seems especially impressive. The clumsy Soviet ASAT may be able to reach altitudes of 1,400 miles,[47] whereas the more accurate US ASAT has an altitude that may be under 500 miles.[48] At the moment ASATs do not realistically threaten anything. The Congressional Office of Technology Assessment, however, estimated in 1985 that: 'if ASAT development received very large resources over a period of time, ASAT technologies would outstrip the countermeasures, and all satellites would become more or less vulnerable'.[49] Since the Soviet ASAT 'threat' has become part of the mixed rationales for SDI, ASAT technology itself has been given a considerable boost in the last three years.

More importantly, many of the technologies of BMD in the SDI are, *ipso facto*, ASAT technologies. Any weapon system—either directed energy or kinetic energy—that comes near being able to hit a missile booster or a re-entry vehicle would have a far easier task hitting a satellite. The target acquisition and tracking systems that would be struggling to identify warheads would have no difficulty in identifying satellites. And all the exotic technologies of SDI which would be struggling to be activated in time to counter a missile attack would have more than enough time to be activated in an ASAT role. The overlap in the technologies of BMD and ASAT, therefore, means that if *anything* effective comes out of the SDI, it will almost certainly be ASAT technology. Even if all BMD technologies are deemed to have failed, it is inevitable that ASAT research, at least, will have been advanced.[50]

A limp argument that is increasingly being expressed in this respect is that it may be possible, with SDI breakthroughs, to use satellites to defend satellites: to harden and defend the C^3 system, in space. In one breathtaking leap General Graham asserts: 'The multiple satellite deployment permits one satellite to defend itself and several others from hostile attack. . . . This system could start deployment in as little as three years and be fully developed in five or six years at a cost of $10–15 billion.'[51] But this simply takes the old offence/defence argument and puts it into space. Even Edward Teller has no doubt that satellites are vulnerable[52] and, as Richard DeLauer testified to the House Armed Services Committee, 'any defensive system can be overcome with proliferation and decoys, decoys, decoys, decoys'.[53] In short there is no way that the US can have BMD without having ASAT, and no way of having ASAT without the Soviet Union eventually having some sort of comparable ASAT.

This fact is made more dramatic and ironic by the present safety of most US satellites in geosynchronous or high orbits. Even a far more mature Soviet ASAT would not be able to threaten US satellites quickly and dramatically. They are simply too far above a ground-launched system and getting an ASAT into a co-orbit takes time. Only three things can really threaten these US satellites: space-*based* (as opposed to ground-based) weaponry, large nuclear explosions in space or the upper atmosphere beyond any reliable hardening capacity, and a determined attack on the infrastructure of the satellite—its data communications, control communications or its ground stations. SDI encourages all three possibilities. At least half the technological challenge of SDI is to develop space-based systems, and if X-ray lasers are to be a part of it then we will have to cross the threshold of accepting nuclear detonations in space. And if the Soviet Union finds itself lagging behind on ASAT, it will always have the option of falling back on the dangerous, easy, third solution: to black the system out and isolate it from the ground. Once both superpowers have passed a fairly moderate threshold of ASAT capability the technological lead which the US will probably have will be almost meaningless.

Finally, we should integrate with these technical developments an outline of the political context. Since 1980 the Reagan Administration has been extremely wary of getting involved in any ASAT limitations.[54] They wanted to keep their own ASAT programme moving and, since ASAT development was only minimally restricted by the 1972 ABM Treaty,[55] it provides an invaluable loophole for the development of the common technology of ASAT and BMD. As the Union of Concerned Scientists expressed it, 'virtually any putative BMD system will be an effective ASAT long before it achieves any significant ABM capability, and the claim that it is really only an ASAT may be difficult to refute for many years'.[56] The relative restraint with which both superpowers approached the militarisation of space up to the late 1970s was based on two general conditions: a combination of technical limitations and a tacit recognition of a degree of mutual interest.[57] The SDI programme has been a major cause (and partly an effect) of a change in both those conditions.

And how will the Soviet Union react? Again, the analysis looks bad for the future of C^3. Stephen Meyer concludes that Soviet responses will be both 'offsetting' and 'emulating'.[58] He bases this on an analysis of Soviet perceptions of the SDI challenge as military, technological and also social. It would, in any event, be a sensible response from the Soviet point of view. Their emulative response is unlikely to be technically sophisticated, so their offsetting response would also be prudent: to develop the capacity to overwhelm any BMD system and to attack the most vulnerable part of it, the C^3.

This brings us to the final illogicality. Faced with an apparently intractable problem of C^3 vulnerability, an inevitable growth in ASAT, and an almost inevitable growth in Soviet targeting of C^3, the SDI aspires to a BMD programme in which a vast, revolutionary C^3 system forms the centrepiece.

SDI AND BATTLE MANAGEMENT

There is widespread agreement that the development of adequate C^3 to manage a BMD battle is of crucial importance to the programme.[59] Indeed, it has been remarked that the computing requirements for the SDI are the only really scientific challenge. The hardware does not yet exist and may never do so: similarly the software. Everything else is essentially a matter of 'heroic engineering'—scaling up and deploying in space existing technologies.

Certainly, the C^3 requirements are almost beyond imagination.[60] To begin with sensors would have to distinguish the 10,000–20,000 warheads within a 'threat-cloud' of 250,000 to 500,000 objects and decoys. Lasers and particle beams may help to distinguish between them. But the system must be capable of deciding how much attention to pay to each object even if it can determine its character. It requires 'cradle-to-grave' tracking information whereby it is known where each object came from, which booster, when it was released, what else has released with it and how much the booster recoiled. It needs a life history of each object. This information has to be constantly updated and distributed to multiple nodes, many of which may be going down simultaneously. So the network would have to reconfigure itself constantly, accurately and without interrupting the flow of information. It would have to avoid the brittleness of complex systems, where a general and catastrophic failure can result from a situation beyond the system's understanding. And it would have to avoid this common pitfall in the face of an adversary who would be deliberately trying to create it. Once target acquisition has been achieved, battle orders have to be issued efficiently, weapons have to be fired and results assessed. The status of one's own forces has to be constantly updated: how much ammunition is left, which weapons are still available, which are optimally placed? And a thin line has to be trodden between covering all legitimate targets and guarding against overkilling—wasting time and ammunition. Faults have to be instantly isolated and compensated for. Enemy spoofing has to be recognised and handled, and some coherent 'human' picture has to be formed for the national command authority to make the political decision as to what to do next.

There is understandable scepticism as to whether the software and the algorithms for such requirements could be created.[61] Something between 40 and 100 million lines of software code would have to be written. And it would have to operate in a 'fault-tolerant' way first time, without previous operational testing. Existing techniques to prove the correctness—without testing—of software cannot yet handle 500-line programs.[62] To handle a 100 million lines would represent an improvement in performance by a factor of 200,000.

The necessary hardware would have to include computers capable of a billion calculations per second. Photonic computers—using light as the source of connectivity between cells rather than electrons—may be capable of this if they turn out to be scientifically possible.[63] Almost all BMD weapon systems will require some configuration of satellites in order to operate. Richard Garwin estimates that satellite constellations, which presently number satellites in tens or at the extreme in hundreds, would have to measure satellite numbers in the thousands.[64] Depending on what assumptions one makes about the threat, the number of satellite battle stations that would be required varies from around 50 to over 2000. And all of these have to be part of the communications network.

Whether or not one is capable of fitting all this into the imagination, certain essential points become clear. Firstly, any move down this road is certain to increase the delicacy and potential vulnerability of strategic C^3 systems. Secondly, it increases the relative importance of C^3 as a front-line element in the strategic balance. And while ground-based systems can be backed up and duplicated by satellites, the reverse is not the case. Any attack on space-based C^3 would have to be the precursor to a general all-out attack. Certainly the defender would be unlikely to stand by while C^3 assets were destroyed, in the hope that it was only a limited attack. Thirdly, the C^3 system itself is likely to become a hair-trigger mechanism. Even existing US C^3 has taken some significant steps towards the predelegation of nuclear launch authority, so short is the time for reaction.[65] A BMD system would have to begin reacting at the instant of an enemy missile launch. Its first decisions on whether to strike at enemy missiles would have to be taken well within the first minute of the action. Within five minutes the system would have to be at full capacity, striking out at the threat cloud. None of these developments are stabilising.

CONCLUSION

Thinking about strategic C^3 is never good for the nerves. In this case it may provoke straight intellectual breakdown. The SDI cannot be so counterproductive to American interests, so self defeating, so wrong headed. In any case it will not happen; the real programme will be a shadow of the original grand design?

A programme can be so self defeating, however, because it goes through so many stages. It begins with a sincere Presidential statement, little critical assessment and an almost immediate confusion of policy goals. This is followed by a downgrading of alternatives and practical problems. Next comes an exaggeration of the threat and a search for new rationales. Finally the programme becomes enmeshed in a series of tactical issues; it is a bargaining chip, an investment in technology, merely a research exercise.[66]

Will the SDI, as such, happen? Almost certainly it will not if we mean will it produce a cost-effective BMD system.

Something, however, is likely to emerge from the programme and, whatever it is, it will have implications for C^3. Any move towards more space-based C^3 is likely to be a leveller between the superpowers over the medium to long term. In itself, this could be a good thing since effective C^3 can be a force for stability. If the levelling process, however, is built on the emergence of ASAT, then the whole C^3 situation is immediately fraught. Strategic stability and crisis stability are then likely to suffer.

It is impossible to say how far down this alarming road we may have gone. But the fact that SDI has set off on this road has already incurred political costs. The time to conclude an ASAT arms control agreement seems to have passed; the previous, informal, regime of restraint has been abandoned and the excitement of the competition and the challenge does not encourage policy makers to consider the centrality of C^3 problems to their future security. Paul Stares reckons that: 'In hindsight, the early part of the 1980s will most probably be viewed as a fundamental watershed in the militarisation of space.'[67] Yes.

NOTES

1. C^3 is also sometimes expressed as C^3I: command, control, communication and intelligence.
2. Good examples include J. D. Steinbruner, 'Nuclear decapitation', *Foreign Policy* **45** (Winter, 1981–2) pp. 16–28; P. Bracken, *The Command and Control of Nuclear Forces*, (Yale University Press, New Haven and London, 1983); B. G. Blair *Strategic Command and Control*, (The Brookings Institution, Washington DC, 1985); D. Ball, *Can Nuclear War Be Controlled?* Adelphi Paper 169, (International Institute for Strategic Studies, London, 1981).
3. 'On balance US national security is served better by an absence of legal restraint upon military competition in space than it would be by an ASAT arms control regime of very uncertain verifiability.' Colin S. Gray, 'Space is not a sanctuary', *Survival* **25**(5) (September/October 1983) p. 196.
4. MIRVs are multiple independently targetable re-entry vehicles. One booster rocket can carry several warheads and each warhead is very accurate. *See* H. F. York, *The Origins of MIRV*, (Taylor and Francis, Stockholm International Peace Research Institute, London, 1973).

5. Geodetic satellites are concerned to map the precise size and shape of the earth's surface and measure the variations in gravity. This information is vital for very accurate targeting.
6. Blair *op . cit.* pp. 212–16.
7. A *geostationary* orbit is one at a sufficient altitude that it takes a satellite twenty-four hours to complete an orbit. It therefore appears to stay in the same place. To do this, however, it must also orbit in the equatorial plane. For the superpowers, however, this is often too far south. They more generally employ a *geosynchronous* orbit which is at the same altitude but at an angle to the equatorial plane. Such an orbit will keep a satellite at the same longtitude but it will appear to dip and rise on the horizon as it crosses and recrosses the equator with each orbit.
8. The other satellite systems are SDS (Satellite Data System) in a polar orbit, NATO III to link with the European allies and the British Skynet LES (Lincoln Experimental Satellite) to link up other satellite communications, and LEASAT which is a series of channels leased from 4 Hughes Corporation commercial satellites. AFSATCOM units ride on FLTSATCOM satellites (for equatorial coverage) SDS (for polar coverage) DSCSIII and LES 8 and 9 satellites as well as other commercial and classified military satellites.
9. The others are, TRANSIT, NOVA (to be incorporated into TRANSIT), NTS and NAVSTAR GPS (Global Positioning System). NAVSTAR is a revolutionary navigational system which will probably become an all-purpose satellite navigational facility.
10. T. Karas, *The New High Ground*, (Simon and Shuster, New York, 1983) p. 112.
11. Union of Concerned Scientists, *The Fallacy of Star Wars*, (Vintage Books, New York, 1984) p. 190.
12. The other satellite systems are, COM 1, COM 2, Molniya 1, Molniya 3, Raduga, Gorizont. *See also* P. Stares, 'U.S. and Soviet military space programs: a comparative assessment', *Daedalus* **114**(2) (Spring 1985) p. 137. *See also* S. M. Meyer, 'Soviet military programmes and the "new high ground" ', *Survival* **25**(5) (September/October 1983) pp. 204–15.
13. S. M. Meyer, 'Soviet strategic programmes and the US SDI', *Survival* **27**(6) (November/December 1985) pp. 276, 285; Union of Concerned Scientists *op. cit.* p. 194.
14. An elliptical, or *Molniya*, orbit is one that brings the satellite low over a certain part of the earth before it then swings to a considerable height as it moves to the outer curve of its ellipse. *Molniya* orbits are therefore variable and the satellite can only spend a certain proportion of its time at an optimum altitude.
15. Blair *op. cit.* pp. 327–31.
16. Ball *op. cit.* p. 20.
17. D. Ball, 'Nuclear war at sea', *International Security* **10**(3) Winter 1985–86) p. 22.
18. P. T. Tarpgaard, *Strategic Command, Control and Communication: Alternative Approaches for Modernization*, (Congressional Budget Office, Washington DC, 1981) p. 13.

19. Blair *op. cit.* p. 182.
20. *ibid.* p. 227.
21. Bracken *op. cit.* pp. 217–18.
22. Blair *op. cit.* p. 182.
23. No one seems to know whether or not TACAMO is an acronym. Either way it has come to stand for 'Take Control and Move Out'. For a detailed analysis of the airborne command systems of the superpowers *see* W. M. Arkin and R. Fieldhouse, 'Nuclear weapon command, control and communications', SIPRI, *World Armaments and Disarmament: SIPRI Yearbook 1984*, (Taylor and Francis, London, 1984) pp. 463–7.
24. After 1987 the present TACAMO fleet of aircraft will be modernised, being hardened against EMP and capable of in-flight refuelling for the first time.
25. Blair *op. cit.* p. 193.
26. The best accounts can be found in Ball, *Can Nuclear War Be Controlled? op. cit.*; R. P. Berman, and J. C. Baker, *Soviet Strategic Forces, Requirements and Responses*, The Brookings Institution, Washington DC, 1982): Arkin and Fieldhouse *op. cit.*
27. Arkin and Fieldhouse *op. cit.* p. 489.
28. *Soviet Strategic Defense Programs* (United States Department of Defense, Department of State, Washington DC, October 1985) pp. 9–10.
29. Ball, *Can Nuclear War be Controlled?*, *op. cit.* p. 44.
30. *See*, for example B. J. Smernoff, 'A bold two-track strategy for space: entering the second quarter-century', in T. L. Heyns (ed.) *Understanding US Strategy: A Reader*, (National Defense University Press, Washington DC, 1983) pp. 119–33.
31. Karas *op. cit.* pp. 66–70.
32. Stares *op. cit.* pp. 139, 141. *See also* his *Space Weapons and US Strategy: Origins and Development* (Croom Helm, London, 1985) p. 252.
33. MILSTAR is the Military Strategic–Tactical and Relay System. *See* Arkin and Fieldhouse *op. cit.* p. 482; Blair *op. cit.* pp. 244–8, 260, 263, 267, 276–7; Karas *op. cit.* pp. 91–2.

    At present, NCA, SAC and NORAD (North American Aerospace Defense Command) mobile command centres have been developed, though their effectiveness and survivability are not known. The JCS and the regional C-in-Cs are in the process of developing regional command centres.
34. Ball 'Nuclear war at sea' *op. cit.* p. 18.
35. Arkin and Fieldhouse *op. cit.* p. 482.
36. Blair *op. cit.* pp. 277–9.
37. Quoted by George Rathjens in a statement to the US Senate Foreign Relations Committee, *Hearings Before the Committee on Foreign Relations*, US Senate, 98th Congress, 1st Session, 14 April, 18 May, 1983, p. 21.
38. Stares, 'US and soviet military space programs' *op. cit.* p. 138.
39. Karas *op. cit.* pp. 137–8.

40. *See* Notes 1 and 23 above.
41. *See* Meyer, 'Soviet military programmes and the "new high ground" ' *op. cit.*
42. Arkin and Fieldhouse *op. cit.* pp. 483–8.
43. Quoted in R. Garwin, 'Star Wars: shield or threat?', *Journal of International Affairs* **39**(1) (1985) p. 38n.
44. Ball, '*Can Nuclear War be Controlled?*' *op. cit.* pp. 35–6.
45. *See* Stares, *Space Weapons and US Strategy: Origins and Development op. cit.* pp. 206–20.
46. On US and Soviet ASATS *see* Union of Concerned Scientists *op. cit.* pp. 198–9, 208–9. *See also* the UCS Briefing Paper, *Antisatellite Weapons* (Cambridge Mass., August 1985). See also Lapraik in this volume, pp. 62–63.
47. UCS Briefing Paper *ibid.* p. 2.
48. D. C. Morrison, 'Defense focus', *National Journal* **18**(3) (18 January 1986) p. 188.
49. IISS, *Strategic Survey 1985–1986*, (International Institute for Strategic Studies, London, 1986) p. 33.
50. On the relationship between the technologies of ASAT and BMD see B. Jasani and C. Lee, *Countdown to Space War* (SIPRI, Taylor and Francis, London, 1984) pp. 58–70.
51. D. O. Graham, 'The high frontier study: a summary' in Heyns *op. cit.* p. 105. *See also* his *High Frontier: A New National Strategy*, (The Heritage Foundation, Washington DC, 1982).
52. R. Garwin, and A. D. White, 'The technical question: can "it" be done?' Paper presented to the World Affairs Council, Philadelphia (4 October 1985) p. 4.
53. H. A. Bethe, *et al.*, 'Space-based ballistic-missile defense', *Scientific American* **251**(4) (October 1984) p. 47.
54. Stares, *Space Weapons and US Strategy: Origins and Development op. cit* pp. 229–35.
55. The only limitation was on ASAT development which threatened those satellites concerned with the verification of strategic arms control agreements.
56. UCS, *The Fallacy of Star Wars op. cit.* p. 225.
57. Stares, *Space Weapons and US Strategy: Origins and Development op. cit.* pp. 237–40.
58. Meyer, 'Soviet strategic programmes and US SDI' *op. cit.* p. 275.
59. *See* R. R. Ropelewski, 'Experts claim battle management paramount to missile defense', *Aviation Week and Space Technology* **124**(2) (13 January 1986) pp. 22–3.
60. Most of the following analysis owes a good deal to The Institute of Electrical and Electronics Engineers Inc., *Spectrum* (September 1985) pp. 33–64.
61. D. Sanger, 'Can software be designed?' *New York Times* (23 October 1985) p. D1. The algorithms refer to the procedures and criteria by which the software decides what it is looking at when it collates information about incoming warheads. They are inherent in the design of the

software. It is necessary, for instance, for the system to have preprogrammed criteria to distinguish between a detected launch that bears all the hallmarks of a standard test flight, a civilian launch, or an attack. Algorithms are as vital as the software package and must be similarly free of major or persistent errors. They must be similarly invulnerable to an opponent's counter-measures, though it is impossible to be certain what those counter-measures might be, and they must be constantly updated to a very high degree of accuracy as the opponent's day-to-day behaviour in space evolves.

62. IEEE, *Spectrum op. cit.* p. 43. Some existing defence systems are built on enormous amounts of software code. The Safeguard Anti-Ballistic Missile System that was abandoned on grounds of its inadequacy, had two million lines of code. More significantly the present World-Wide Military Command and Control System, which is the US's overall C^3 infrastructure, contains ten million lines of code. There have been numerous and persistent errors in those portions of it that are activated regularly. In 1977 a test of the system revealed that a staggering 62 per cent of messages failed to reach their destinations. *See* J. Lamb, 'The bugs in the Star Wars programme', *New Scientist* **108** (21 November 1985) p. 28.
63. W. Broad, 'Computers of the future may run on light, not electricity', *International Herald Tribune* (24 October 1985) p. 8.
64. IEEE *Spectrum op. cit.* p. 41; Garwin and White *op. cit.* p. 3. *See also* R. Garwin, 'How many orbiting lasers for boost-phase intercept?', *Nature* **315** (23 May 1985) pp. 286–90.
65. Arkin and Fieldhouse *op. cit.* p. 501.
66. This list derives from a succession of stages suggested in a lecture by Richard Garwin.
67. Stares, *Space Weapons and US Strategy: Origins and Development op. cit.* p. 235.

3 The Technology of Weapons in Space—Past, Present and Near Term*

Squadron Leader Robert Lapraik

Since the dawn of the space age, military planners have been acutely aware of the benefits of utilising the 'new high ground' and placing weapons in space. The evolution of ballistic missile defence (BMD) is certain to have a growing impact on strategic offensive capabilities and the development of anti satellite (ASAT) weaponry will increasingly affect strategic stability during time of peace or political crisis. This chapter aims to provide a review of the technology that has been harnessed for both offensive and defensive space weapons over the last two and a half decades. The discussion of past systems serves to highlight some of the difficulties inherent in ground-based BMD and ASAT systems, many of which would also be relevant to future technologies. Additionally, the capability of present and near-term technologies is assessed and the interrelationship between the BMD and ASAT roles is examined in the light of existing arms control agreements.

In the early 1960s the prospect of building a constellation of orbital bombs appeared to offer significant advantages over the rudimentary ICBMs of that era. They would remain in orbit—a modern sword of Damocles—to be called into action against any target on Earth with little warning. However, a brief examination of this apparently simple concept reveals a number of severe operational limitations.[1] Some of these limitations, for example the requirement for a very accurate deboost, are unique to an orbital bombardment system,

* The views expressed in this chapter are those of the author and do not represent the position of the RAF.

whereas others, such as limited availability (sometimes called absenteeism) and the difficulty of in-orbit maintenance, are common features of most space-based weapons.

The area of influence of a space-based weapon is largely determined by the associated suborbital ground track. This ground track is a function of orbital dynamics, specifically the interrelationship of the motion of the weapon around its orbit and the daily rotation of the Earth about its axis. Since the orbital plane of the weapon or battle station essentially remains fixed in direction in space, whilst the Earth rotates beneath it, the ground track appears as a sine curve with the maximum excursions north and south of the equator approximating to the angle of inclination of the orbit.[2] Altitude, on the other hand, determines orbital period and lifetime. The period varies from ninety minutes at 200 km altitude to the synchronous, twenty-four-hour orbit at 36,000 km. Selection of an operational altitude for a weapon system will therefore be determined by such considerations as effective weapon range and the frequency of revisit over a particular point on the Earth's surface.

If an orbital bombardment system were to have 'whole world coverage' then it would have to be placed into a near 90°, polar orbit. The targets that would be vulnerable during each orbit would lie inside a swath centred on the satellite ground track, the width of the swath being a measure of the manoeuvre capability of the re-entry vehicle (RV).

A consequence of space basing the weapon, in this case an orbital bomb, is that it would only threaten its selected target twice per day, when the target came under, or at least close to, the orbital plane. For this reason an operational system, capable of attack at any time, would require a large constellation of similar but carefully spaced orbits, accepting that most of the weapons would be unavailable at the instant of an attack. Rather than being a simple concept, the orbital bomb therefore emerges as being quite complex and inefficient. In practice there would be other penalties: each system would have to compromise a power source, altitude control and propulsion subsystems and these would have to function reliably after many years in space. Precise control at deboost would be essential for acceptable accuracy, since the dynamics of re-entry are intolerant of errors in the quantity, direction and timing of the deboosting rocket burn. In addition to the problem of accuracy, this system would be further complicated by the general difficulty of long-term serviceability and in-orbit maintenance.

Both superpowers acknowledge the problems of accuracy, reliability and availability and on 27 January 1967 they signed, together with sixty-three other nations, the Outer Space Treaty which prohibits 'the placing into orbit around the Earth of any objects carrying nuclear weapons or any other kinds of weapons of mass destruction.[3] This signalled the end of orbital nuclear bombs; however, the ink was barely dry when it was discovered that the Soviet Union was developing another space weapon known as Fractional Orbit Bombardment System (FOBS).[4] Here the warhead was to be placed into a low earth orbit, rather than a traditional ballistic lob. As well as providing unlimited range, its operational advantage over the ICBM rested on its low orbit (around 150 km), compared with a typical ICBM trajectory (1,200 km at apogee) and hence its much later penetration of early warning radars. Not only did FOBS provide less warning of an attack but also, launched south-east from the Soviet Union, the warhead would travel across the Indian Ocean and approach the United States through the then unprotected 'backdoor'. The Soviet Union completed seventeen FOBS tests over the period September 1966 to August 1971 and deployed a number of operational FOBS launchers to their missile launch centre at Tyuratam. However, the utility of this delivery technique was limited by similar accuracy problems to those of the orbital bomb. It would be launched into low earth orbit to pass over the target during its first orbit and would again require precise deboosting. Additionally, once space-based early-warning satellites became operational the major advantage of surprise had been lost.[5] The two superpowers apparently found no difficulty in agreeing to ban FOBS under the terms of the unratified, but generally adhered to, SALT II Treaty.[6] the deployment of nuclear weapons into Earth orbit, either as a means of delivery or space basing, had, for the time being at least, been relegated to the history book. The ICBM, therefore, remained the primary means of delivering strategic nuclear weapons.

Since the advent of operational ballistic missiles, planners sought an effective defence. The forerunner of US BMD systems was the Nike-Zeus,[7] a variant of the Nike-Hercules antiaircraft missile. This was typical of first-generation BMD systems. A powerful tracking radar was used to acquire the target RV and determine its precise trajectory. Once launched, the Nike-Zeus was itself monitored by a missile-tracking radar and both sets of data were fed to a master computer for the calculation of intercept geometry and appropriate detonation of the nuclear warhead. The RV would be destroyed outside the atmosphere by the massive dose of X-rays absorbed in its structure. This US Army project began in February 1955 and started a fierce debate on the desirability and technical feasibility of BMD, particularly against a background of limited funding and in the light of possible advances in ICBM technology. One advance that could pose an effective counter to this system was the deployment of a number of lightweight decoys to confuse the target tracking radar. The effect of these decoys was to magnify the apparent threat, or 'threat cloud', and hide the real warheads within that cloud. However, the counter-counter-measure was already within sight in the form of the Nike-X ABM system, which was configured to intercept RVs after they had re-entered the atmosphere.[8] Here the aerodynamic heating and the ionised wake of hot gases increase the radar-echoing area and these, together with the difficulty of matching the RV trajectory once within the atmosphere, provide the ability to discriminate real warheads from decoys. This late interception, however, required the use of sophisticated phased-array radars and imposed severe performance parameters on the interceptor missile and control system. Despite advice from the science establishment to the contrary, in September 1967 Secretary of Defense McNamara announced a decision to go ahead with the Nike-X area BMD, by then named Sentinel. This offered a 'thin' defence against the perceived Chinese nuclear threat, together with protection from accidental launch and an option to defend Minuteman ICBM silos. Nevertheless, a combination of public concern over the siting, near major cities, and high cost led the new Nixon Administration to suspend Sentinel in February 1969, in favour of a revised ABM programme, Safeguard, now to provide point defence for ICBMs.[9] Safeguard came under intense political pressure but narrowly survived on the basis of the Soviet BMD effort around Moscow and the continuing growth of the Soviet strategic offensive arsenal. Like Nike-Zeus before it, Safeguard employed two types of radar, the missile site radar (MSR) and the perimeter acquisition radar (PAR), but now two missiles were used: the long-range exoatmospheric Spartan and the short-range endoatmospheric Sprint. Safeguard was therefore the first example of BMD architecture involving two 'layers' of defence. The PAR was a large phased-array radar designed to acquire target RVs at long range. This information was used to define the RV's trajectory and predict impact point. The shorter-range MSR provided precise target data for the final interception, under the control of a high speed computer. Spartan was a three-stage, solid-propellant missile, launched from an underground silo and carrying a five-megaton command fused nuclear warhead. It is estimated that this large warhead would produce a lethal radius in space of 5 to 10 km. Sprint, on the other hand, was a small, high acceleration, two-stage missile which was ejected from an underground silo by a gas-propelled piston before its solid rocket motor ignited, providing an acceleration of around 100 g. Since Sprint was designed for defence at altitudes around 15,000 metres, a much smaller nuclear warhead of the order of ten kilotons was employed, destroying the RV by a combination of blast and high energy neutrons. Although Safeguard had the capability to provide both area defence, by intercepting outside the atmosphere, and point defence, using Sprint in the terminal phase, its specific objective, as announced by President Nixon on 14 March 1969, was to protect the US deterrent system and so deny the Soviet Union the opportunity to acquire a pre-emptive counter-force capability. Notwithstanding its 'layered' configuration, Safeguard PAR and MSR radars would have been vulnerable to the use of decoys and chaff in midcourse and premature detonation of incoming RVs (or even interception itself) would have degraded radar performance enough to allow penetration by subsequent RVs.[10]

Early Soviet BMD research and development (R&D) took

place at Sary Shagan, conveniently positioned about 1,000 miles due east of the Soviet Union's first ballistic missile test centre at Kapustin Yar. Early development was revealed by U-2 reconnaissance photographs in April 1960. However, the next mission in May ended when Gary Powers was shot down and so this type of overflight reconnaissance ceased; fortunately the era of surveillance from space was just beginning, and so further monitoring of this and other BMD development was carried out by photoreconnaissance satellites. The first Soviet missile deployment believed to have had a BMD capability was that around Leningrad. By 1962 some thirty launch positions for Griffon antiaircraft missiles were under construction around the city.[11] The fact that they were already being dismantled in 1964 indicated that the Soviets were unhappy with Griffon's limited performance; nevertheless, the concept of utilising a long-range antiaircraft missile in the BMD role is a design philosophy that the Soviets pursue to this day.

The most important Soviet BMD effort took place around Moscow. This system used large, powerful radars enabling interceptions to take place at long range using the Galosh ABM, the Soviet equivalent of Spartan. Galosh was specifically designed for long-range exoatmospheric interception; measuring some eighteen metres in height it had a range of around 300 km and used a multimegaton nuclear warhead to achieve its kill. A network of very large Hen House radars was constructed around the periphery of the Soviet Union, providing both early-warning and target acquisition for the Moscow defences. Dog House, and later Cat House, phased-array radars, positioned to the South of Moscow, provided the battle management function. Each complete installation comprised two target-tracking, four missile-tracking radars and launch facilities for sixteen missiles. By the spring of 1970 four Galosh complexes had been completed, comprising a total of sixty-four missiles.[12] However, despite the use of phased-array radars and the ability to intercept at long range, the Moscow system would only have provided a token defence against a concerted American attack. The defences would have been swamped by large numbers, particularly if MIRVing were employed and the radars were vulnerable to the use of decoys and were unhardened against both direct and indirect nuclear effects. The fact that the Soviets limited their deployment to sixty-four missiles probably reflects an acceptance of the limited operational capability of their BMD technology at that time.

When the SALT I talks began in Helsinki in late 1969, both superpowers therefore had a very limited BMD capability. The two missile elements of Safeguard, Spartan and Sprint were undergoing testing at the Kwajalein range, whilst Galosh was entering service around Moscow. Nevertheless, the Soviets were aware of the relative superiority of the US in this field; by early 1972 some twenty-three Spartan and Sprint test firings had taken place from Kwajalein, of which nineteen were classed as being successful.[13] It is therefore probable that the Soviet Union's willingness to negotiate ABM limitations derived from a growing concern over the disparity in US/USSR BMD potential. The 1972 ABM Treaty limited both sides to two missile defence systems, one to protect an ICBM launch site and the other to defend the national capital. In 1974, a protocol to the treaty limited each side to only one ABM site of not more than 100 missiles; the United States continued its Safeguard deployment at Grand Forks, whilst the Soviet Union retained the Moscow system. The Soviets had therefore achieved their objective of limiting the effect of the US BMD technological advantage.

In the event, the $6 billion Safeguard system was declared operational on 1 October 1975, only to be closed down by Congressional decree the following day. From that point, the US interest in BMD R&D went into decline, but this was not the case in the Soviet Union. The early 1980s saw the enhancement of the Moscow defences with the replacement of the original sixty-four above ground, Galosh launchers with silo-based versions of improved performance. Also, with the introduction of a high acceleration missile, designated SH-8, for interceptions within the atmosphere, the Soviets opted for a second layer of defence and gained the operational advantage of atmospheric discrimination between RVs and decoys. This update is expected to be complete by 1987,[14] bringing the Moscow defence up to the 100 launchers permitted by the 1972 ABM Treaty.

In addition to this modernisation of the Moscow defence, the US believes that the Soviet Union is developing a rapidly deployable ABM system.[15] Such a system would probably comprise mobile phased-array radars and a combination of new SH-4 exoatmospheric interceptors and SH-8, high acceleration missiles, for terminal defence. It has been suggested that such a system could be integrated with new phased-array radars under construction at six locations within the USSR, to provide a nationwide missile defence. In line with previous Soviet design philosophy, such a deployment could be augmented by any BMD capability possessed by deployed surface-to-air missiles. In this respect both the SA-10 and SA-12 are believed to have a BMD capability. Although more suited to the trajectory of a tactical ballistic missile or SLBM, they may also provide a valuable potential against strategic missiles which could be exploited in a widespread ABM deployment.[16]

The flight of a ballistic missile may usefully be divided into four distinct phases: the boost phase, whilst the booster is thrusting; the postboost phase, while individual MIRVs and decoys are being dispensed from the 'bus'; the midcourse phase, whilst RVs and decoys follow a ballistic flight path through space; and the terminal phase, whilst the RVs descend through the atmosphere to their targets. Having reviewed the technology employed by BMD systems to date, certain observations can be made. ABMs, setting aside the secondary capability of some surface-to-air missiles (SAMs), have been expensive, purpose-built missiles that relied on vulnerable, ground-based radars for acquisition and command guidance to the target. In general, a fairly large nuclear warhead has been employed to compensate for errors in guidance. It is clear that, within these constraints, RVs have only been vulnerable to interception during the last part of the midcourse phase and all, bar the final portion, of the terminal phase. It is equally clear that interceptions in the boost or postboost phases can only be achieved from space, whether by space basing the weapon itself or relaying the energy of a ground-based laser via mirrors placed in space. However, new technologies are, even now, emerging to satisfy the demands of interception in the midcourse and terminal phases.

An example of this new technology was ably demonstrated in June 1984 when the US Army carried out the fourth test of its Homing Overlay Experiment (HOE). The target, a dummy warhead launched by a Minuteman ICBM from Vandenberg Air Force Base (AFB), was successfully intercepted by a kinetic energy weapon (KEW) launched from Kwajalein.[17] The KEW carried a Honeywell long-wavelength infra red (IR) sensor to provide target acquisition and terminal homing. The device was launched by a modified Minuteman, after the incoming RV had been acquired and tracked by a radar in the Kwajalein atoll. The interception resulted in a direct hit, head on, at over 160 km altitude, and serves to emphasise the validity of the concept of non-nuclear kill (NNK) using present technology. Although the HOE programme is now complete, it has surely paved the way for both endoatmospheric and exoatmospheric KEWs in the near term. It also indicates an important area of overlap between the technologies of the BMD and ASAT roles. Owing to the convergence of these technologies, it would seem appropriate to review ASAT technology to the present, before moving on to consider prospects for the near term.

A satellite in earth orbit might appear to be a relatively simple target. It is, after all, a very delicate piece of equipment, powered by large solar arrays, it maintains orientation in space using optical devices, and, as yet, possesses no self-defence armament. On the other hand, even the lowest of satellites is far away, at least 200 km, is moving very fast, of the order of 8 km/s, and measures only a few metres across.

The techniques used to place an ASAT into the vicinity of its target can usefully be divided into two categories. The first is known as the 'direct ascent' method. Here, the interceptor may be launched from any position on earth, and at any time, provided that the target is within range. However, using this profile, the initial boost into orbit is critical, since any error during this phase could require unacceptable manoeuvring by the ASAT during the final stages of the intercept. The ASAT will always be at high speed relative to its target and the geometry of the terminal phase will necessitate a very capable homing system. Alternatively, a large warhead would have to be employed to compensate for terminal inaccuracies.

The United States has concentrated almost exclusively on

this direct ascent method of attack. Responding to a possible threat from Soviet orbital bombs, the US Army developed their first operational ASAT, an adaptation of the Nike-Zeus ABM.[18] The programme was approved by McNamara in May 1962 and was deployed to Kwajalein. The system was operational in 1964 and provided an 'interim' ASAT capability to counter Soviet low orbit satellites. It had a fast reaction time and used a nuclear warhead to achieve its 'kill'. In the same period, the US Air Force was proceeding with another direct ascent ASAT based on the liquid-fuelled Thor missile. Despite a longer reaction time, the Thor had the operational advantage of a greater altitude capability of around 1200 km. This system was based on Johnson Island and was declared operational in May 1964. The Nike-Zeus ASAT was phased out from 1966 and the Thor system was finally withdrawn in 1975. In addition to the utilisation of the direct ascent method, two other points of interest emerge: both the Nike-Zeus and Thor ASATS used a nuclear warhead to widen their lethal radius and since 1975 the US has had no operational ASAT capability.

The effect of detonating a nuclear warhead in space became clear in July 1962 when the US carried out a high altitude nuclear weapons test, codenamed Starfish.[19] In this test a 1.4 megaton device was detonated at about 400 km over the Pacific and as a result a number of orbiting satellites were fatally damaged by the effects of SGEMP. The test also had a widespread effect on ground-based equipment, causing extensive power failures in Hawaii, many hundreds of kilometres from the detonation. Nuclear tests in the atmosphere or in space were subsequently banned under the terms of the Limited Test Ban Treaty of 1963; nevertheless, the experience of Starfish was a powerful pointer to the problems that would result from the detonation of nuclear-tipped ASATS or ABMs in near-Earth space. The detonations would take place over, or at least close to, friendly territory, causing damage to unprotected electronic components on the ground, in the air and in space. Satellites are difficult to harden against EMP as the electrons causing the pulse within the satellite are dislodged from the shielding material itself. The prospects of scoring an 'own goal' and damaging friendly satellite resources are very real and so Western planners have, where possible, moved away from the use of nuclear-tipped weapons in space.[20]

The next development of the direct ascent method of attack was also American and was spurred by continued Soviet ASAT activity. The programme, initially on a small scale, was formally given approval in 1977 by the outgoing Ford Administration. It is based on a miniature homing vehicle (MHV) developed by the Vought Corporation, as a result of previous research into ABM technology.[21] The device is very small, measuring only some 30 by 33 cm. It is spin stabilised in space and has no moving parts. Detection of the target satellite is achieved by eight IR telescopes focused on to a cooled IR focal plane array. Lateral manoeuvre is achieved by fifty-six one-shot solid rocket boosters girdling the middle of the vehicle and controlled by a sophisticated ring laser gyroscope, integrated with an advanced microcomputer. The weapon weighs 15 kg and travels at around 50,000 k.p.h., achieving its kill by impact. The MHV represents impressive technology, but the major advantage of the system derives from its novel boost concept. The ASAT is launched from an F-15 in flight, using a relatively small two-stage solid composite as a booster. The first stage is based on the Boeing SRAM, whilst the second stage has an Altair III rocket motor developed from the fourth stage of the Scout launch vehicle. The total package weighs around 1,200 kg, measures 5.2 metres in length and occupies the centreline weapons station of the F-15, 'Celestial' Eagle. Prior to take-off the F-15 is programmed with precise target parameters and the full mission profile from ASAT Mission Control Centre located in NORAD's Cheyenne mountain complex. When commanded, by on-board software, the pilot zoom climbs to around 12,000 metres and initiates ASAT launch.[22] The ASAT climbs under inertial guidance to orbital altitude and then homes to impact, using its IR sensor.

The F-15 ASAT therefore enjoys the inherent flexibility of the direct ascent method of attack, whilst also having the mobility associated with an airborne launch platform. However, the F-15 ASAT is limited in altitude capability to less than 1,000 km and requires the mobility of its launch vehicle to achieve a near head-on intercept.[23] Many important Soviet satellites, for example those communications or early warning satellites in highly elliptical 'Molniya' orbits, would prove difficult targets and would require F-15 deployment well outside the continental US. The most likely targets of an operational F-15 ASAT would seem to be the Soviet reconnaissance, ELINT and radar satellites. However, the $4 billion programme is some way from becoming operational and, although the first test against a target in space was successfully concluded with the destruction of P78-1 (Solwind) on 13 September 1985, further progress is dependent on continued support from the US Congress.

The second technique of ASAT attack is the technically more forgiving, but tactically less flexible. This is the coplanar interception and is the method employed by the Soviets. Here, the ASAT is launched into the same orbital plane as its target; the only practical way to achieve this is to initiate ASAT launch when the launch site comes under the orbital plane of the target, and this occurs only twice per day. Having launched into the plane of the target then only the phasing, or relative position in the orbit, remains to be solved. Phasing can be adjusted relatively easily by varying the orbital altitude; however, in practice this would have to be completed within, at the most, two orbits of the ASAT, otherwise the enemy would have ample time to track the ASAT, determine its target and manoeuvre the victim satellite out of reach.

The Soviet Union has consistently employed the coplanar technique and began testing its ASAT, against satellite targets, in 1968. The ASAT operates from launch sites at Tyuratam and employs a conventional fragmentation warhead with a lethal radius of about 1 km. In all, since 1968, twenty tests are believed to have been carried out, of which nine are considered to have been successful in achieving a miss distance of less than 1 km.[24] The tests fall into two distinct series, employing different technology and interception techniques. The first, from October 1968 to November 1971, used active radar for the terminal homing, interception taking place after two orbits of the ASAT. A break of over four years preceded the second series of tests, from February 1976 to June 1982. In some of these interceptions the radar homing device was retained but the attack was completed within one orbit of the ASAT. The technique placed the ASAT into a low orbit, where it enjoyed a high speed, quickly overhauling its target, to 'pop-up' for the final attack manoeuvre. These tests achieved a success rate of about 50 per cent and would have the tactical advantage of providing little warning of attack. In addition, this second series of tests also saw the use of an optical homing system where the ASAT attempted to home on to the thermal emissions of its target; however, of the six attempts there were no credited successes.

Although the Soviet ASAT launch vehicle, the SL-11 (a modified SS-9 ICBM), is estimated to have the capability of lofting the ASAT to 5,000 km altitude, the highest successful interception occurred at just under 1,600 km. Also, all ASAT tests have taken place in the limited inclination band of 62° to 66°. Within this height and inclination band the Soviets can only attack the US NOSS (National Oceanographic Surveillance System) and the Chinese reconnaissance satellites. This led to speculation that Soviet concern over the Chinese space programme played a major part in the formulation of the Soviet ASAT effort. Nevertheless, of those satellites within range of the ASAT, the US 'Polar' orbiters such as the Transit Navigation Satellites, the military meteorological satellites and the US photosurveillance satellites (Big Bird and KH-11) would be prime Soviet targets. However, the fact that the Soviets have never tested their ASAT at these angles of inclination casts some doubt on the true operational and, indeed, intended capability of the system.

The current Soviet ASAT is therefore a reasonably reliable but rudimentary system. The radar homing device would be susceptible to jamming or deception and would be unable to counter timely target evasion. The booster is a large liquid-fuelled vehicle, which is slow to prepare and launch; the launch sites are few and easily targeted. Above all, though, its limited altitude capability leaves those satellites in highly elliptical, semisynchronous or geosynchronous orbit completely immune from attack. The homing head could possibly be placed on top of a larger launcher (for example the SL-12) to increase its performance, but this would become an extremely costly venture and there is nothing to suggest that the Soviets intend to follow this route. Perhaps they are content with their current limited ASAT performance whilst they look to other, more exotic, technology for future ASAT use.

Before looking to the future, it is worth discussing two further ASAT techniques that are within the scope of present technology. The first is rarely mentioned and involves the use of electronic warfare (EW) to counter or degrade satellite operations. Few satellites currently in orbit possess any electronic counter-counter-measures (ECCM), and would therefore be vulnerable to the effects of high-powered uplink jamming being directed into their antennas. Such jamming could interfere with communications, or data links, and more subtle techniques could tamper with the operation of the many satellite subsystems. Unless precautions are taken, an adversary could even take over control of the satellite for his own use. Not only is EW a cost-effective ASAT weapon but there is no reason why it should not be employed against those satellites in the crucial geostationary orbit currently immune from physical attack. However, now that these space assets are assuming such importance to the military, steps are being taken to introduce the degree of protection required to counter the EW threat perceived over the lifetime of a satellite. This includes the use of phased-array antenna to minimise the effects of uplink jamming and advanced digital processing, and to reduce the likelihood of interception or interference with communications and data links; nevertheless, advances in technology are bound to perpetuate the vicious circle of measure, counter-measure and counter-counter-measure in this electronic battle.

The second, apparently simple, method of causing physical damage to a satellite is the use of space mines. Here the explosive device would be parked in orbit, within lethal range of its target, to be detonated at will. The satellite could be dedicated to this purpose or the explosive could form a package on an otherwise operational satellite. However, if the victim satellite possesses a manoeuvre capability, which most do, then this would clearly have to be matched by the space mine. Also, with the increasing capability of radars and optical systems to 'see' into space, the mine would have to possess a high degree of 'stealth' if it were to escape detection. It therefore emerges that the space mine would not, in fact, be a simple or cheap device, that to be viable the mine would have to be considerably cheaper than its target, and that the close 'shadowing' of the victim satellite by the mine could be seen as being extremely provocative.

Of those weapons likely to become available in the near term, many would be capable of a dual BMD or ASAT role. Some could be developed and tested as BMD components within the terms of the ABM Treaty, but others, contravening the treaty, could equally well be developed as ASAT weapons, in the first instance, to be scaled up for BMD later on.

In the near term, two technologies stand out for consideration: kinetic energy weapons employing non-nuclear kill and ground-based lasers (GBLs). In the first instance KEWs could comprise conventional chemical rockets, similar to Spartan and Sprint, to provide a BMD capability both within and above the atmosphere. Terminal guidance would most likely be of the IR type, similar to that used in the HOE or MHV. As non-nuclear interceptors, flight tests from fixed ground-based launchers could take place within the terms of the ABM Treaty, at agreed test ranges, providing the launchers lacked the capability of being rapidly reloaded or of launching more than one interceptor missile.[25] These low acceleration KEWs would be supplemented later by electromagnetic launchers (EMLs). Using EMLs, much higher acceleration is possible, leading to velocities much greater than those attainable with chemical rockets. EMLs broadly break down into two main types: the mass driver and the railgun.[26] The former is an AC linear motor which accelerates the projectile using a travelling magnetic wave, capable of velocities of the order of hundreds of metres per second. The latter, on the other hand, is essentially a linear DC motor, utilising the Lorentz force to accelerate the projectile at some 100,000 *g* to perhaps tens of kilometres per second. In the railgun high currents, of the order of several mega-amps, provide a large acceleration, and therefore high velocity, over a relatively short distance. Already in the laboratory a 318 g projectile has been accelerated to 4.2 km/s over a distance of 5 metres. Other lighter particles have reached velocities of over 10 km/s. The current challenge is to scale such a device to cope with projectiles of around 1 kg and accelerate them at around 200,000 *g* to velocities of 20 to 30 km/s. in addition, the projectiles would need to be guided, using either IR terminal homing or some form of beam riding (for example, laser communications link to the rear of the projectile for guidance information). In order to withstand the very high acceleration, the device would have to be entirely solid state and would require substantial propellant manoeuvrability during the terminal phase of the attack. If such a weapon can be produced, it could be tested with both guided and unguided projectiles, provided it is not tested 'in an ABM mode'. Inevitably, railguns of this performance would ideally be space based to provide important kinetic energy layers of defence during the boost, postboost and midcourse phases of flight. Testing railguns in space could be achieved against 'satellite' targets in the ASAT role, so effectively sidestepping the terms of the ABM Treaty. Despite their relatively slow speed and long time of flight to the target, KEWs have the advantage that they would complicate the provision of effective countermeasures and, additionally, kill assessment would be relatively straight forward.

Of the other more exotic technologies that may one day be harnessed for BMD, that of the high energy laser (HEL) is the most mature. The first HEL deployment is likely to be ground based, in order to overcome the problem of supplying the high power necessary to drive the laser. A ground-based laser would also enjoy a stable aiming platform; however, the atmosphere presents a medium that is virtually opaque to the radiation emitted by many lasers. Any GBL would be constrained to operate at a wavelength corresponding to a 'window' in the atmosphere, at which there is little atmospheric absorption. Another effect of passing the laser through the atmosphere is the distortion caused by atmospheric turbulence. For weapons applications, this distortion would have to be counteracted by introducing appropriate compensating distortion into the emitted beam, such that the beam arrives undistorted at the target.[27] This compensation requires the use of complex, adaptive optics and presupposes that the atmospheric distortion can be monitored. Of course such a laser would be defeated by cloud, dust or other atmospheric pollution and so the preferred location is likely to be on a mountain top, and yet still convenient for the provision of many megawatts of power.

The damage mechanism of a HEL depends on the nature of the target's surface and the type of laser employed. The proportion of incident energy absorbed by the target is dependent on the degree of coupling of the beam to the surface. Using a continuous high-powered laser the surface material vapourises, forming a plasma. The plasma acts as a frequency shifter and re-emits the laser energy it absorbs, at shorter wavelengths. This enhances coupling of the laser energy to the target and thermal damage occurs from the surface structure inwards. A pulsed laser, on the other hand, capitalises on the explosive generation of the plasma at the surface. This causes an impulsive loading to be felt within the surface material and it is this that is exploited in a pulsed laser, to store plastic energy in the target until surface fracture occurs. The pulsed laser, therefore, employs both thermal weakening and mechanical damage, providing a more efficient use of available energy than thermal melt-through alone.[28]

The capacity of a laser to damage ICBMs is largely dependent on the laser 'brightness'. The brighter the laser, the more energy it is able to deposit on the target. Brightness, and therefore weapon effectiveness, increases with increasing beam power, increasing mirror diameter and decreasing laser wavelength.[29] Neither power nor optics size should prove to be limiting factors for GBLs; however, decreasing the wavelength would require an ever-increasing quality of optical components. Also, since a beam of high energy density is likely to have little beam divergence, it would require very accurate pointing and tracking (typically of the order of one microradian).

Currently the brightest laser in the West is the US Navy's MIRACL (midinfrared advanced chemical laser). MIRACL is a 2.2 MW deuterium fluoride (DF) laser that uses decay of DF molecules to lase at 3.8 μm. This wavelength corresponds to an atmospheric 'window' and has good transmission properties. Designed for US Navy investigation into shipborne defence, MIRACL was deployed to the high energy laser systems test facility at White Sand, New Mexico, in 1984 and is being used for vulnerability and lethality testing.[30] During such a test, on 6 September 1985, MIRACL, combined with the Sealite beam director, destroyed the second stage of a Titan 1 missile that was pressurised to simulate flight loads. This test dramatically demonstrated the potential vulnerability of ballistic missile boosters; however, powerful though MIRACL is, it should be noted that for practical BMD purposes an increase in laser brightness of three to four orders of magnitude would be required.

In addition to increasing output power, an important step towards increasing brightness would be to reduce the wavelength by about an order of magnitude to the range 0.1 to 0.4 μm. So, although chemical lasers are the most mature, research is now turning towards those operating at shorter wavelengths; the excimer and free-electron lasers. Excimer lasers operate by the electrical stimulation of a mixture of inert gases, such as in the Los Alamos krypton fluoride laser. However, these lasers tend to be large, heavy and fairly inefficient. The laser currently attracting most interest for the ground-based role is the free-electron laser (FEL). In the FEL a high energy electron beam interacts with a magnetic 'wiggler' field to convert kinetic energy from the electron beam into laser radiation. Importantly, the wavelength of the laser can be tuned over a wide range by changing the dimensions of the magnetic 'wiggler' field.

The most promising near-term GBL option would therefore appear to be high-powered FELs deployed on high ground, clear of atmospheric pollution and located in areas of good weather. Adaptive optics would be required to predistort the emitted beam to compensate for its passage through the atmosphere. Essentially these GBLs would provide point defences, although their versatility could be extended by the use of space-based mirrors, to relay the laser energy on to RV targets, whilst in the midcourse. However, such a constellation of space mirrors represents a severe technical challenge. Such mirrors would have to be light, yet rugged, structures of high optical quality and capable of very accurate pointing and tracking. Unless the mirrors were placed at geostationary height (which could impose low elevation angles), the absentee problem would necessitate a large constellation of relay and fighting mirrors, all linked by a complex C^3I network. This is unlikely to be accomplished in the near term.

Inevitably GBLs will be tested against targets in space, to

perfect the propagation and correction techniques. This testing need not be carried out against strategic missiles and can therefore be accomplished within the terms of the ABM Treaty. In the first instance, therefore, it is likely that the satellite will become the primary target of the GBL. Not only would such a laser make an excellent ASAT device, but also the fact that the ABM Treaty places no restraint on the development, testing or deployment of antisatellite technology would, of itself, drive ASAT laser technology as a first step towards a BMD capability.

Not many years from now, high energy GBLs may prove capable of causing damage to the structure of orbiting satellites through heat and shock action. In practice, though, the vulnerability of a satellite to a particular laser would depend largely on the altitude of its orbit, since altitude combines with laser beam divergence, to determine the energy density that would be deposited on the satellite's surface. However, a satellite is vulnerable to laser energy at much lower levels than that required for structural damage.[31] Optical and IR sensors could be blinded by laser illumination within their field of view and permanent damage could result from the secondary effects of heat or electrical overloading. These sensors could be the 'eyes' of surveillance satellites, or the horizon sensors, used by many operational types to maintain correct orientation with the Earth. In addition, most satellites maintain a delicate heat balance in order to keep their electronics within defined temperature limits; even low-powered laser illumination could upset this balance and render the satellite useless, simply by raising its temperature.

Of course the technology that can produce high-powered GBLs can also spawn an array of counter-measures.[32] Targets could be polished to reflect laser energy or covered with an ablative material; the simple act of spinning the device would serve to dissipate the incident energy. For the military satellite, stealth and laser hardening will become standard design requirements. Optical filters and laser detectors will protect sensitive optics from damage by laser weapons and increasing use will be made of on-board (nuclear) power, to avoid the necessity for large, vulnerable solar panels. The era of equipping satellites with ASAT counter-measures has begun and will no doubt reach its logical conclusion and provide active defence in due course.

The current strategy of deterrence is being exposed to increasing pressure by advances in ICBM technology, whereby highly accurate counter-force missiles are beginning to favour the 'first strike'. Near-term BMD technology offers the prospect of a stabilising influence, by providing some measure of protection to the retaliatory forces, and so preserving the regime of mutually assured destruction. In the near term these defences are likely to comprise both KEWs and GBLs constrained, like their predecessors, to intercept RVs during the late midcourse and re-entry phases of ICBM flight. Whether or not these defences prove effective against incoming RVs, the prospects of survival for vital satellite systems seems to be heavily dependent on sophisticated ASAT countermeasures. Whether, on balance, this will make the world a safer place over the next few years remains to be seen.

NOTES

1. Curtis Peebles, *Battle for Space* (Blandford Press, 1983) p. 50.
2. Inclination is the angle between the orbital plane of the satellite and the equatorial plane of the Earth.
3. *World Armaments and Disarmament, SIPRI Yearbook 1973* (Paul Elek Ltd, London 1973) Appendix 13B.
4. Curtis Peebles, *Battle for Space*, p. 64.
5. Under the American Defence Support Program, satellites using IR sensors detect and report the launch of ballistic missiles to NORAD, located in the Cheyenne Mountain Complex.
6. *World Armaments and Disarmament, SIPRI Yearbook 1984* (Taylor and Francis, London 1984), Ch. 18.
7. Ashton B. Carter and David N. Schwartz, *Ballistic Missile Defense*, (Brookings Institution, Washington DC, 1984) Ch. 9.
8. *ibid.* p. 336.
9. J. W. R. Taylor (ed.) *Jane's All The World's Aircraft 1972–73* (Sampson, Low, Marston and Co Ltd, London, 1972) p. 589.
10. Ashton B. Carter and David N. Schwartz, *Ballistic Missile Defense*, p. 77.
11. *ibid.* p. 195.
12. J. W. R. Taylor (ed.) *Jane's All The World's Aircraft 1972–73*, p. 594.
13. *ibid.* p. 589.

14. Department of Defense, *Soviet Military Power 1985*, p. 48.
15. *ibid.*
16. *ibid.*
17. Hans A. Bethe, Richard L. Garwin, Kurt Gottfried and Henry W. Kendall, 'Space-based ballistic missile defense', *Scientific American* (October 1984) p. 45.
18. Curtis Peebles, *Battle for Space*, p. 82.
19. John A. Adam and Mark Fischetti, 'SDI: the ground experiment', *IEEE Spectrum* (September 1985) p. 45.
20. Thomas Karas, *The New High Ground* (Simon and Schuster, New York, 1983) p. 149.
21. *ibid.* p. 151.
22. Keith F. Mordoff, 'Test ASAT launched autonomously from USAF F-15 carrier aircraft', *Aviation Week and Space Technology* (7 October 1985) p. 18.
23. James E. Oberg, 'A dozen anti-ASAT fallacies', *Air Force Magazine* (July 1985) p. 47.
24. Richard L. Garwin, Kurt Gottfried and Donald L. Hafner, 'Antisatellite weapons', *Scientific American* (June 1984) p. 29.
25. United States Information Office, 'SDI research in compliance with the ABM Treaty', *Official Text* (23 April 1985).
26. Arthur Knoth, 'Electromagnetic acceleration—a new ballistics', *Military Technology* (6/84).
27. Lori McClelland, 'The bright outlook for lasers', *Defence Electronics* (July 1985).
28. Kosta Tsipis in William Gutteridge and Trevor Taylor (eds.) *The Dangers of New Weapon Systems* (Macmillan Press, London, 1983) p. 27.
29. 'SDI: the grand experiment', *IEEE Spectrum* (September 1985) p. 50.
30. Michael A. Dornheim, 'Missile destroyed in first SDI test at high-energy laser facility', *Aviation Week and Space Technology* (23 September 1985) p. 17.
31. Jeff Hecht, *Beam Weapons The Next Arms Race* (Plenum Press, London, 1984) Ch. 12.
32. *ibid.* Ch. 9.

4 Problems and Prospects of Strategic Defence

Philip Webber

THE OVERALL CONCEPT

SDI-I and SDI-II

The SDI concept introduced into the public eye by President Reagan in his now (in)famous 'Star Wars' speech of 23 March 1983 is of 'an inpenetrable shield' which would render nuclear weapons 'impotent and obsolete'. The idea is that this nuclear 'umbrella' would stop any incoming missiles, thus rendering the world free of the menace of nuclear devastation.

The idea itself is not that new; President Reagan was to some extent reiterating old ideas about ABM systems which had not been thought seriously about since 1975, when the US abandoned the Safeguard system at Grand Forks Air Force Base in North Dakota. The system was assessed to be not worth its maintenance costs and to provide inadequate defence.

What was new about Presidents Reagan's speech, however, was the incredible scope of his proposal, now labelled 'ballistic missile defence' (BMD). No longer would ABM systems protect a few US missile fields against possible surprise attack by the Soviets, but now, the system would cover the entire United States with a 'peace shield'; what is more, once the technology had been developed, the Americans would share this technology with the Soviets, thus ensuring peace for the world both East and West.

The above 'umbrella' version of the strategic defence concept can be called SDI-I. Nevertheless, it is *not* the concept being actively promoted under the 'Star Wars' or 'high frontier' banner. Current proposals do not seek to protect the entire US and Europe, but to concentrate upon protection of 'hard' military targets such as missile silos.

Little more than a year after President Reagan's speech, Lt. Gen. James Abrahamson was chosen by Weinberger, the US Defence Secretary, as Director of the Strategic Defense Initiative Office *or* (SDIO). Spending of $25 billion is envisaged for the first phase,[1] while the further three phases would take up to twenty years to complete. The phases are the research phase (up to 1990), the systems development phase (prototype development and hardware testing), the transition phase (possibly involving applying leverage upon the Soviets to 'build down' nuclear weapons) and the final phase (deployment of a multiphase system).

Details of the proposed SDI system itself now began to emerge. The system was seen as destroying oncoming missiles in several defensive 'layers' (see Figure 1). A missile would have to survive passing through all the layers before it could reach its intended target. However, none of the proponents of this multilayered version of SDI are now claiming that such a scheme could stop *all* missiles, which is necessary for President Reagan's original vision of the future. General Abrahamson, for example, has talked of three to seven layers, each 50 to 80 per cent effective.[2] These estimates are not based upon any firm technical assessment, but for three 80 per cent effective layers it would mean that *at best* about 120 warheads out of an attacking force of 12,000 would penetrate the shield, destroying up to 120 US cities. The new version of SDI, which we can call SDI-II, is now not intended to protect cities and people but to enhance the 'survivability' of missiles in their silos, to 'enhance' the effectiveness of deterrence and to persuade the Soviets of the wisdom of 'building down' or reducing the worldwide stockpiles of nuclear weapons. With the notable exception of the President, no one in the US administration is now arguing in favour of an inpenetrable peace shield. The SDIO, however, clearly benefits from the favourable publicity generated in the public eye by SDI-I.

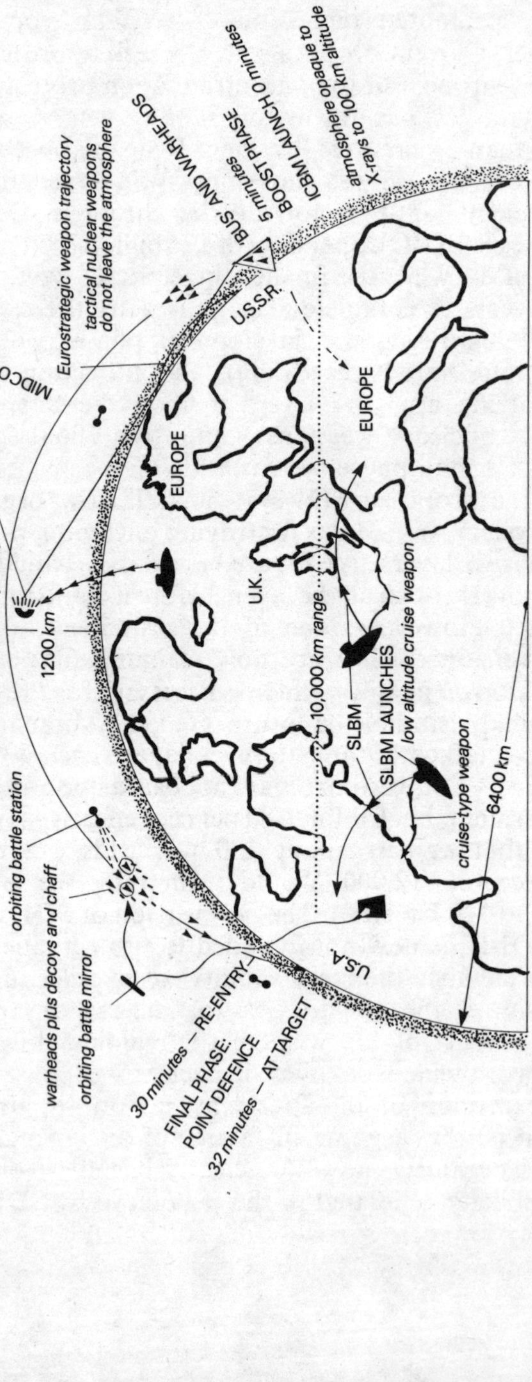

Figure 1. The atmosphere is a very thin layer (less than 150 km) of highly absorbent and turbulent gas, absorbing most types of laser beam and destroying any particle beam or 'smart rock' in a short distance due to friction. The U.S. and the U.S.S.R. are separated by approximately 90°, therefore attacks from the U.S. on the Soviet boost phase must either come from space or from mirrors in space.

THE BASIC CONCEPTS

The flight of an intercontinental ballistic missile begins with its launch, known as the 'boost' phase, during which time the separate stages of the delivery vehicle burn propellants to generate enough energy to throw the postboost phase vehicle or 'bus' containing several warheads to the point in space where the warheads separate and go to their individual targets. This is the basis of the MIRV warhead dispensing system. Each warhead 're-enters' the Earth's atmosphere from space separately. Typical US and Soviet MIRVs contain ten separate warheads plus many more decoys. During the boost phase, intensely hot exhaust gases are produced in large quantities, providing a very clear heat signal and radar reflection which can be detected easily from satellites in orbit or over-the-horizon early-warning radars. During this phase the bus is carried to altitudes of 100 to 200 km in a few minutes. After this time the fuel burn is over and the bus and warheads plunge into space and give off no signals for easy detection by sensors.

In the postboost phase, the bus is finely adjusted in its ballistic flight for positional or launch errors and deploys its warheads in a preprogrammed sequence. At the same time a cloud of decoys, 'chaff' (radar-reflecting metal strips and metal-coated balloons and other more sophisticated electronic devices) are also deployed with the intention of confusing or saturating warning radars or other sensor systems designed to track their trajectories.

In the third, midcourse phase, lasting ten to twenty minutes depending on the distance between launch site and the target, the warheads reach altitudes of 1,200 to 1,600 km, travelling within their cloud of decoys, some of which can generate several radar copies of themselves and make ten warheads look like more than a hundred. In this portion of their flight the warheads are very difficult to distinguish from their decoys.

In the final or re-entry phase the warheads re-encounter the Earth's atmosphere and are slowed down by atmospheric drag. The warheads are designed to penetrate the atmosphere without damage and finally to correct their courses to hit their targets on the ground or at sea. Most of the decoys would at this stage be 'swept' away by the atmosphere or burn up. Less than one minute is left before they reach their targets.

From the point of view of a proponent of SDI, all these four phases have their advantages and disadvantages.

During the boost phase the missile is travelling largely through the atmosphere and, in the case of intercontinental range missiles, over enemy territory. It is, however, very hot and produces strong infrared and near-infrared emission. In the postboost phase the bus can receive signals from navigation satellites but would emit no signals for detection. A proposed new generation of sophisticated sensors might, however, find and track the object. After the decoy cloud is deployed the warhead is very hard to distinguish and even harder to track directly, although its original boost course would provide quite accurate fixes. There is no atmosphere to protect the warheads in this phase. Upon re-entry the decoys are progressively stripped away by the Earth's atmosphere but the warheads are now closer than 130 or so kilometres and will have to be attacked over home territory where a nuclear detonation, even to stop a warhead, may cause significant damage itself.

The system concepts envisaged so far naturally split up into weapons designed to attack warheads from within the atmosphere (endoatmospheric systems), and systems designed to attack them from outside the atmosphere, (exoatmospheric systems).[3] The specific weapons categories themselves split up into the so-called 'directed energy' weapons—lasers and particle beams, 'kinetic energy' weapons—high velocity impact devices actually to physically hit the warheads (so-called 'smart-rock' technology); and antimissile systems similar to the original ABM systems.[4]

There is a very distinct conceptual split between the *hardware*, which could undoubtedly destroy warheads, and the *systems*, which have the much more difficult task of actually finding, tracking and aiming the weapons. Finally, there is the most difficult task of all, co-ordinating and optimising the entire attack upon the warheads so that the least number get through (battle management).

It should be noted that these SDI concepts do not directly address the specific problems of warheads delivered from submarines close offshore where the flight time is much shorter and the warhead does not have to go into space, warheads delivered by low-flying Cruise missiles or bombers, or nuclear bombs planted in advance in enemy territory or in merchantmen moored in ports. All these possibilities are beyond the scope of proposed SDI systems.

SYSTEM COMPONENTS[5]

Lasers

A laser is a device that produces an extremely intense and pure type of radiation (generally in the visible, infrared (heat) or ultraviolet parts of the spectrum). Because the light is pure, or 'coherent' a laser beam does not diverge as much as an ordinary light beam as it passes through the atmosphere or space. The beam travels at the speed of light, which is practically instantaneous for near-Earth applications. Thus the laser seems an ideal weapon. The problems are in getting enough power into the laser without destroying it in the process and in aiming the entire beam. The lasers under research by DARPA (Defense Project Research Agency) are of two main types: those which already exist and are known to work (albeit at much lower power than needed for SDI) and those which are rather more speculative—such as the proposed X-ray laser.[6] DARPA is attempting to develop a space-based chemical laser system. The ALPHA project aims to develop a multimegawatt laser (one megawatt is the energy output of 1,000 normal 1,000 watt bar heaters) which would be powered by the mixing of two highly reactive gases such as chlorine and fluorine, which would also have to be carried up into space. Calculations show that a 100 megawatt system would need one to two seconds accurately trained upon its target to damage sufficiently a missile booster (assuming a range of 3,000 km, spot size on target of 1 m^2 and two seconds between shots). If the system has three minutes to attack 1,500 ICBMs, and assuming two seconds retargeting time, roughly thirty lasers would be needed for targeting. Because of the curvature of the Earth and the fact that the lasers would constantly be orbiting the Earth, not all lasers could be in line of sight, thus increasing the necessary number. The required number of battle stations has been the source of much controversy, and numbers as high as 2,300 have been produced.[7] There is now general agreement between opponents and proponents of SDI that around 100 stations would be needed under optimum projected conditions to attack all of a Soviet ICBM force.[8] The problem caused by the curvature of the Earth can be avoided to some extent by preferential targeting of lasers to targets at shorter range. Several hundred shuttle loads would be needed for the fuel alone. Many other flights would be required to construct the laser stations.

An alternative to putting these highly vulnerable (and potentially highly explosive) laser systems in orbit is to base lasers such as a 'free-electron' or 'excimer' laser on the ground and to reflect the laser beams from mirrors in space. The excimer laser creates high energy pulses from excited molecules at ultraviolet wavelengths, while the free-electron laser works by passing a beam of electrons through a specially shaped magnetic field. There are several problems with space-based mirror systems, however. To avoid beam divergence the mirrors would have to be over ten metres across—larger than any precision mirror constructed before—and of a highly reflective material to avoid being destroyed by the laser beam itself. The mirrors themselves would also be highly vulnerable to attack. The only way round this problem is to mount the mirrors in a 'pop-up' mode, where they would be launched into space only when attack was detected. The problem then is just how to get them high enough quickly enough for the laser beam to get round the curve of the Earth to hit attacking missiles in the boost phase. New very high energy boost rocket fuel would have to be invented and in advance of that used by the Soviets otherwise the mirrors could not possibly accelerate faster than their targets, which is absolutely vital if vital minutes are not to be lost.

The most controversial SDI proposal of all is the so-called X-ray laser.[9] This concept involves using a small nuclear explosion in space to generate an intense but brief burst of X-ray radiation from laser tubes surrounding the bomb. The X-ray laser would be a one-shot' system, as it would destroy itself as it was fired, therefore many laser rods, probably comprising fine metal fibres, are envisaged surrounding the central nuclear device pointing in various directions towards ICBM targets.

This system would at least be lighter than the huge space mirrors or chemical laser systems; also an X-ray laser beam would be highly destructive. It would not, however,penetrate the atmosphere, so it would have to be deployed into and fired from space. To have a chance of 'popping up' far enough and quickly enough its proponents envisage its deployment in submarines off the Soviet coastline.

Lawrence Livermore Laboratory tested some possible components for a nuclear-powered laser in a top-secret test in the Nevada desert in March 1985. There followed a dispute over whether the test was a success or not. Sources both from within Livermore itself and from Los Alamos (another US weapons development lab) showed that the data were faulty. Despite this the lab's Deputy Associate Director George Miller used the erroneous data in a report to the Pentagon in April to justify further funding.[10]

Particle beams

Particle beams are a highly energetic stream of particles which can be charged, such as electrons or ions or electrically neutral particles. Such a beam would deal a blow to a target rather than burning its surface as in the case of a laser. Such systems at present exist only in highly specialised and very expensive research facilities involving the generation of extremely high voltages and superconducting magnets such as in CERN, where minute amounts of energy can at present be generated in a beam after great effort. Beams of charged particles would, moreover, spread out as they passed through space due to the mutual repulsion of the particles making them up. Also, all charged beams are bent by magnetic fields and the Earth is a huge magnet with a field that is constantly and unpredictably varying. Finally, devices to generate neutral particle beams would weigh several tons and be even more costly and expensive than the laser systems.

Hit and kill devices

These would be devices literally to throw projectiles of some sort at ICBM warheads, destroying or disabling them by sheer impact. For the quickest response times, such devices would again have to deployed in space to avoid problems with air resistance. Interceptor-borne devices based upon a small rocket launched from a high-flying jet have successfully destroyed high-flying missile warheads in tests. However, they are not contenders for destroying missiles in their boost phase because they are not quick enough. In space a railgun could fire high speed projectiles, but only over relatively short ranges because the projectile speed would be too low to hit targets at longer ranges. The system would thus have to be placed in low orbit, meaning that hundreds of systems would be required for any sort of global coverage. Because of all these factors, these devices make most sense as part of a re-entry phase defence or as part of a point defence systems around a missile silo.

COMMAND CONTROL AND TARGETING

Command, control and targeting is the most important component of any putative SDI system. No matter how clever or complex the solutions to actually destroying threat warheads in the various stages of their flight, the means to destroy them have to be co-ordinated in a highly complex battle management system. The problems such a battlefield management system has to cope with are immense. The systems would have to:

1. Detect and respond to any potential launch threat anywhere in the world from the Soviet land mass or Eastern Europe to most of the oceans.
2. Evaluate the threat and respond in an appropriate way within seconds of launch by deploying and activating the optimum combination of defensive systems to have the best chance of destroying the missile or missiles.
3. Aim and fire the chosen weapons systems at the target, check after firing to see if it is still a threat and then fire again if it has not been successful.
4. Distinguish quickly between peaceful Soviet launches and missile flight tests.
5. Once past the initial boost phase, be able to cope with tracking, evaluating and targeting a vast number of objects in space. According to the Pentagon Defense

Technologies Study Team, the Star Wars system might have to track up to 300,000 objects such as warheads, fragments of warheads, and decoys.[11]

6. Predict the courses of all these objects, distinguish them from its own space systems and try to distinguish the warheads from the decoys.
7. Be hardened against its own generation of blinding pulses of nuclear electromagnetic radiation, when weapons systems such as the nuclear-powered X-ray laser are being used, quite apart from any hostile countermeasures.
8. Use computer software written for this task that would need to make decisions without human intervention (there would be not enough time for humans to make political decisions or judgements) based upon preprogrammed 'rules of engagement'. It would also need to be a suite of programs far larger and far more complex than any previously written set of programs.

There is a very significant programming difficulty in clearly defining the overall problem which the programmes would be expected to solve in the first place. It is rather like having to design a world-wide airline reservation system which has to work first time when it isn't known what sort of planes you have (how many seats, etc.), nor to where and from where people may wish to travel nor even *when* they may want to travel. The airports and numbers of terminals are not even clear.

Estimates vary,[12] but probably about one hundred million lines of programming would be necessary to contain the system, which would have to process roughly sixty megabytes of information relating to the objects in space. The listing would be a staggering three million pages long, which would have to be written and checked for mistakes and faultlessly loaded on to the battle-management system computers. These computers would have to be built with sufficient redundancy to guarantee that they would work without crashing during attack, or indeed under attack themselves.

The sheer scale of the programming effort would necessitate the use of computers to perform some of the programming, but the most difficult and crucial stage of the programming effort would be in attempting to specify and program for all the possible eventualities and combinations of events which the system might have to cope with. This is always the hardest part of the programmer's effort, when a flow diagram has to be constructed of how the entire system is supposed to work and how the various subsystems are supposed to interact and convey information to each other. How should the system react to a Soviet missile aimed at a target in East Germany, for example? What is it to do about French, British or Chinese missiles?

Thus program design is very difficult. Moreover, no one human being could keep the whole program definition in their head and the overall concept would have to be defined by whole teams of programmers. Concerning the next stage of the programming, implementation, present computers are simply not fast enough to perform the likely computations quickly enough without massive parallel processing being necessary (where several computers are run simultaneously on the same problem). The final stage (of normal programs) is testing. Here the problems for the SDI software start to look truly mind boggling. Clearly the system could never be tested against a real attack; the system could be partially tested by checking its output when known inputs are presented to it and checking that the answers were those expected. Unfortunately, with a program as complex as this, thc answers, not to mention the test data, would be almost as complicated to understand as the program itself. These answers would also not check the real time of response (the system has to solve certain problems within a time limited by the attack). To do this the system would have to be partially tested against a small simulated attack. This would be very unsatisfactory. In large systems faults often show up only once the system is loaded to near its capacity and it is only at such times that the true operating speed of the system under threat could be evaluated. As more demands are made upon such systems the slower they get. Any basic conceptual mistakes in the programmer's flowcharts could not be very well tested by other programmers; they would probably make similar errors. Another problem is that very complex systems such as this do not always react the same way to the same stimuli. The entire operation system would have billions of possible operating conditions, only a small fraction of which could ever be tested for systematically. Thus even a completely successful test could be no guarantee that the systems would work in a real attack.

Moreover, what do you do if you *do* find an error? Correction of errors is not a trivial problem. Often, correction of one error leads to another. This is why real computer operating systems or software contain 'bugs' which are left uncorrected for fear of creating more damage. Past experience with very expensive, complex computer systems is not reassuring. One launch of the Space Shuttle was delayed because of hitches in communication between just four computers on board. A Star Wars system would need hundreds to thousands of computer subsystems all continuously in touch with each other and continually being updated with information. The Shuttle software amounts to about 500,000 lines of code, much less than the SDI system, and the error leading to flight-time sychronisation error was introduced due to the correction of another error two years previously.[13] It did not reveal itself upon all occasions but, on average, about once in sixty-seven times.

Communication between the SDI computers would have to be via a worldwide communication system. The US (WWMCCS) is an existing communication network used by civilian and military authorities to transmit information to and from US military forces in the field. The WWMCCS performs well until it is heavily loaded. In a large-scale test in 1977 the chances of successful message transmission were reduced to 38 per cent.[14] An SDI systems could not tolerate this.

In the *Aegis* battle-management system designed merely to protect one target the system failed to shoot down six out of sixteen targets because of faulty software. But earlier smaller-scale simulation tests had not discovered programming errors. Even a test relating to the SDI project itself is not great cause for rejoicing. The crew of the Space Shuttle were supposed to position the craft so that a laser beam aimed from a mountain at 10,023 feet should be reflected off a mirror on the side of the shuttle. A simpler experiment is hard to imagine. It failed initially because the computer controlling the Shuttle interpreted the laser source position as 10,023 nautical *miles* above sea level rather than that number of feet.[15] Needless to say this fault was easily rectified and the test was successful a few days later.

Finally, program writing would represent an immense management and security problem. Programming would entail over 30,000 person-years of work, or 3,000 programmers and analysts working over a period of ten years. The inevitably high turnover in projects of this sort always means that program documentation has to be very good to avoid mistakes as new people take over from others. Partly finished programs have to be picked up, understood and all necessary corrections made without missing any. In normal systems, failure to correct one subroutine would be picked up later in testing, but this would not be possible in the SDI system. Similar errors have led to the deaths of aircrews, when they have not been informed that navigation computers have been updated. Such errors could not be tolerated in the Star Wars battle-management system. With such a large number of programmers involved in such a project, security would pose special problems. There would be the risk of Soviet agents infiltrating the team and deliberately inserting hard-to-find errors that would cause the system to fail at some critical time. The very high levels of security that would be needed to try to prevent such problems would mean that communication between programmers would be restricted on a 'need to know' basis. This would increase the likelihood of unintentional errors.

COUNTER-MEASURES AND COUNTER-COUNTER-MEASURES

It is a mistake to evaluate a putative SDI system upon a straightforward extrapolation of present Soviet capabilities. In the history of military technology, all developments in some form of defensive system have been accompanied by parallel developments in offensive measures and counter-measures. Any assessment of the reality or otherwise of a future SDI system must take into account actions which the Soviet Union will or could take in response.

The first point is that if the US has developed means of

shooting down some Soviet missiles, then similar means would be able to shoot down elements of the SDI system itself. Much is made of the fact that a slow-moving multiple warhead missile represents considerable concentration of assets as it takes off in its vulnerable boost phase. What then of the device that is supposed to shoot it down? If a battle station is prepositioned in an orbit in space where it might be required to destroy, say, 100 missiles, destruction of the battle station would represent protection for that same number of weapons. If the battle station is prepositioned, in space then its position can be accurately plotted days in advance. It would be vulnerable to even a few grains of sand placed in an opposing orbit from another orbiting station designed to do this. It could not manouevre out of the way of a laser beam or an oncoming railgun projectile. Overall, destruction of a set of battle stations or battle mirrors in orbit is a simpler problem than the task they themselves have, of destroying the much larger numbers of and far more unpredictable missiles taking off.

Any battle stations permanently in orbit would therefore need self-defence mechanisms to avoid their own early destruction. The most vulnerable part of the system would, however, be the sensors which would be vital to track warhead trajectories. These monitors would be highly vulnerable to intense heat or light from lasers or to nuclear explosions high up in the atmosphere where the generation of the nuclear electromagnetic pulse is at its maximum. These monitoring stations could *not* be deployed in a pop-up mode because twenty-four-hour-all-round coverage would be necessary to warn of attack.

In comparison with the sensor system and battle stations, nuclear launch vehicles are quite robust, particularly after the boost phase. Very simple counter-measures complicate the task of the attacking SDI system. Rotation of the boosters as they take off means that roughly ten times more energy or ten times as long would be needed to burn a hole in the rocket casing. Reflective coatings could increase these figures by a further 100 times. After all, if the SDI system can use a mirror to point its beams, then a rocket booster can use the same technique to reflect them. The rocket boosters could be designed to burn unevenly to fool the satellite sensors, to burn more quickly to give them less time and to perform unpredictable changes of trajectory to prevent preselection of particular targets at an early stage. For submarine launches of shorter range (so-called depressed trajectory) warhead flights can be programmed to avoid the necessity to rise into space, reducing the time that they would be vulnerable to postboost phase attack.

In the postboost phase it is a simple matter to increase the number of decoys, electronic and otherwise. The simplest decoy is a gas-filled reflective balloon. Such decoys can be detected by noting their recoil to a sensing-laser beam, but this complicated the SDI system task severely. Warheads can also be hidden *inside* balloons. Warheads could be designed to appear to be destroyed by emitting clouds of debris under attack, preventing the system from re-identifying them. They would have to be attacked again.

In the near re-entry phase a very simple counter-measure is called 'salvage fusing'. This means that warheads that were attacked after a certain part of their flight would explode in a nuclear explosion anyway. At altitudes below 9,000 metres this would have a severe blast and fire effect upon populated areas, farmland or forests. This last counter-measure is possible because of the extremely fast trigger time of the nuclear detonators used in nuclear weapons, which can operate far faster than any destructive measure can become fully effective. Salvage fusing means that all late re-entry defences would only have any effect in protecting 'hard' targets such as missiles silos where the warhead has to explode at exactly ground level to create the maximum damage. For 'soft' targets such as cities or airports any terminal systems would have little or no defensive effect.

Finally, no SDI system could cope with low-flying Cruise missiles, nuclear weapons smuggled into countries for use in war or nuclear detonations from weapons planted in ships in harbour.

IMPLICATIONS OF TECHNICAL DEVELOPMENTS

The overall assessment of all the technical factors has several implications, some of which pose severe constraints upon any possible SDI system. Firstly, the task of protecting 'hard' point targets in the US is the easiest task. The warning time is longest and not all the oncoming warheads have to be targeted as some would impact on other targets, far simplifying the battlefield management task. Counter-measures would have the least impact upon the final defensive phase in this mode.

The task of successfully defending cities or extended populated areas in the US is a much harder task, because the targets are themselves much more vulnerable. Even if the bulk of the warheads were destroyed, the plutonium released by them could still re-enter over the US, and would be spread over the entire northern hemisphere, creating a severe radiation hazard to life for thousands of years to come. Less than half of 1 per cent of warheads getting through would have catastrophic consequences in any case. But a 99.5 per cent reliable system design is simply not a possibility for many reasons, particularly hardware faults, inability to perform full software testing and the unknown effects of counter-measures. Experience with all complex systems gives no cause for reassurance about the likelihood of advances in the future which will change these simple facts because of the all-pervasive influence of simple human error in design and implementation.

Some proponents of SDI argue that a limited version of the system could easily be applied to the 'tactical' situation in Europe or the Middle East. The argument goes that if SDI can cope with several thousand warheads in an all-out attack then a smaller-scale 'tactical' war would pose far less problems. This is not the case. Although the computational task would be smaller, the system would have to respond in a very short time. In a tactical nuclear conflict, flight times of warheads would be less than one minute in some cases. This incredibly short time would be a very severe test of the system analysis capability. Another problem is that tactical warheads do not even necessarily leave the Earth's atmosphere, posing a problem for space-based killer systems.

The system would also have no way of deciding when conflict in Europe or the Middle East has become nuclear. In a real conflict that suddenly went nuclear, very large amounts of military air traffic would become potential targets. There is no way that the system could know when or if an aircraft was carrying nuclear or non-nuclear weapons. Weapons such as Cruise and 'stand-off' missiles are identical in flight and external appearance in their nuclear and non-nuclear versions. It seems much more likely that if a system could be devised that was powerful enough to hit ground or near-ground targets, that the SDI system would become embroiled in any conventional conflict. This would be highly dangerous for both sides because the computerised system would be just as likely to destroy NATO as Warsaw Treaty planes or weapons. This would be particularly embarrasing if the SDI systems stopped a planned NATO first use of nuclear weapons! If the SDI system was programmed *not* to become involved then it could not be used once the situation had gone nuclear without some command signal to this effect. The SDI planners would seek to avoid such inputs to the system for fear of possible mistakes or false messages. Because of the extremely short times involved there would be no opportunity for the three or four phases of defence. SDI planners talk of one 'lumped' phase. Because of the existence of only one phase, the system effectiveness would be dramatically reduced, quite apart from all the other problems for tactical operations. Warhead salvage fusing would remove any possible protection from European cities, as would depressed trajectories for weapons fired from submarines in the seas around Europe or the Middle East.

What SDI systems *will* be capable of very early on, even if none of the other systems work, is of having a very considerable antisalellite capability. Even a 1 per cent effective SDI system would knock out most satellite early-warning systems. This is because such systems are highly sensitive and thus vulnerable. This is a very strategically destabilising implication of SDI.

SDI, working or not working, will be incredibly expensive. It will not be cost effective. The cost of the destructive SDI systems will be far more than the counter-measures to confound them. The system would be highly vulnerable to attack and will represent a massive concentration of military assets and thus a high priority military target. Much of the system (as are many complex systems deployed on the battlefield today) would have to de designed to protect itself.

This would create a technological bureaucracy of systems which would themselves need protecting. and so on. The overall cost of the system will escalate as a result of such factors.

A partially effective SDI system would be a highly effective defence against nuclear strikes upon the US by a minor nuclear power such as the UK, France or China. Once SDI was installed, the concept of an 'independent' deterrent or one of *tous azimuts* (the French 'all-direction' deterrent) would have no meaning.

SDI is presented as a purely defensive system. Whether this is true or not, depends on what your definitions of 'offensive' and 'defensive' are. The actual components of any SDI system would be necessity by highly destructive weapons of extremely high accuracy and of global reach and range. The SDI 'shield' is not a passive concept similar to the idea of a 'peace shield' of President Reagan, but a much more active and dynamic idea capable of use for defence *or* offence. An operative SDI system would be capable of use as an enhancement of highly offensive overall strategies such as nuclear first strike. If the US possessed a complete SDI system while the Soviets did not, the US SDI system could ensure (as far as it worked) that no large-scale Soviet response would be made to a US disarming first strike upon the bulk of its weapons in land-based missile systems. It is unrealistic and naive to assume that the Soviet Union would see SDI as a purely defensive measure. Thus the very concept of SDI (as President Reagan acknowledged in his original speech) undermines the strategic balance of terror. It would in fact tip the balance markedly towards the US and enhance US options for 'nuclear diplomacy'. However, the Soviet Union will not allow the US to seize the 'high ground' of space unopposed.[16] They will further develop their own SDI and anti-SDI capabilities. In any future superpower conflict, the first shots could well be fired in space between the opposing battle stations and space mines before the nuclear weapons were launched. The Soviets have already categorically stated that they would deploy additional ICBMs and weapons such as cruise to circumvent SDI. In as much as the US fear that the system might not work, the Soviet response will be seen as excessive, thus provoking further US measures. SDI thus provides another new driving force for the arms race. SDI will usher in a new era not of peace but of a new weapons technology race in space with the new emergence of justifiable fears of first strike.

For all the reasons described above, this is not a race into which the US should enter. In view of recent problems with the US Space Shuttle, the failed Titan launch of a KH-11 spy satellite, and the very high Soviet technological competence in space, this may not even be a race which the US will win.

NOTES

1. D. Baker, *Star Wars: The Strategic Defence Initiative* (Lecture to BAAS, Summer 1986) *passim.*
2. Various media sources (Autumn 1985).
3. G. Canavan, H. Flicker, L. Hantel, O. Judd, D. Roeder, K. Taggart and J. Taylor, *Alternative Concepts Evaluation for Strategic Defence* (Los Alamos National Laboratory, New Mexico, 87545, Paper LA-UR-85-2210, 1 July 1985) *passim.*
4. *SDI: Progress and Challenges*, Staff Report submitted to Senators Proxmine, Johnston and Chiles (17 March 1986).
5. S. French, *Star Wars: the Technology and Politics of Star Wars* (SANA, London Production Centre, London, 1985) *passim*; and J. Turman, *The Fallacy of Star Wars* (Vintage Books, New York, 1984) *passim.*
6. L. Allen and N. Dombey, 'X-ray lasers to shoot holes in test ban treaty', *New Scientist* (19 September 1985) *passim.*
7. J. Turman, *op. cit.*
8. G. Canavan, H. Flicker, O. Judd and T. Taggart, *Comparison of Analyses of Strategic Defense* (Los Alamos National Laboratory, New Mexico 87545, Paper LA-UR-85-754, 1 February 1985) *passim*; D. Baker, *op. cit.*, J. Turman, (ed.), *Empty Promise, the Growing Case Against Star Wars* (Beacon, Boston, 1986) *passim.*
9. L. Allen and N. Dombey, *op. cit.*
10. *See Nuclear Times* (January/February 1986) p. 23.
11. P. Marsh, 'Anatomy of star wars', *New Scientist*, (14 November 1985) p. 32.
12. H. Lin, 'The development of software for ballistic missile defense', *Scientific American* (December 1985) p. 32; J. Lamb, 'The bugs in the star wars program', *New Scientist* (21 November 1985) p. 27; D. Baker, *op. cit.*; P. Marsh, *ibid.*
13. H. Lin, *ibid. passim.*
14. J. Lamb, *op. cit. passim.*
15. H. Lin, *op. cit.* p. 37.
16. V. Olenev (Pseudonym), *Detente* (March 1986) *passim.*

Part II
The Superpowers

5 Space and Strategic Defence Reconsidered*

Group Captain Timothy Garden†

From the time that the United States first became vulnerable to possible nuclear attack from Soviet missiles, strategists have agonised over the paradoxes of mutual assured destruction (MAD). During the Sixties, Robert McNamara, the US Secretary for Defense, looked for ways to limit the damage to the American homeland in any future superpower conflict, so as to enhance deterrence. Such measures included counter-force targeting, civil defence and ABM systems. After significant expenditure in all these areas he concluded in 1965, in a statement to the Committee on Armed Services that:

> If we were to try to secure survival of a very high percentage of our population and if the Soviets were to decide to frustrate this attempt because they viewed it as a threat to their assured destruction capability, the extra cost to them would be substantially less than the extra cost to us.[1]

This view was not universally accepted, and work on the ABM systems continued on both sides. While McNamara did not believe it could offer any significant benefit against a massive Soviet attack, he thought protection against the potential Chinese nuclear threat to the US might be feasible. However, in 1966, the Joint Chiefs, backed by a strong Congressional lobby, sought at heavy Soviet-orientated ABM system, which was to be deployed as quickly as possible. The announcement, late that year, of evidence of a Soviet ABM

* This chapter is based upon the author's two articles 'Space and strategic defence—the final frontier', and 'Strategic defence reconsidered', published in *The Hawk* in March 1984 and March 1985 respectively.

† The views expressed in this chapter are the author's own.

system being developed, added weight to the pro-ABM lobby. The technical difficulties, and the very high cost, led to a decision to deploy only a light system in the US. This was described as having a defensive role against the possible Chinese threat to cities, a capability to make the Minuteman sites less vulnerable to Soviet attacks, and as a safety measure against accidental firing of ICBMs against the US. The Sentinel programme, as it was known, soon ran into difficulties. In March 1969, President Nixon announced that a defence against heavy Soviet attack was not feasible, and that the programme needed to be modified. The new Safeguard ABM system was to be designed purely to protect ICBM sites—a much more limited aim. This did not stem the debate, both inside and outside of government. There was concern about the cost of deploying such a limited system, and its rationale finally became that of a bargaining chip for arms limitation. Although we are not privy to the debates, which undoubtedly took place, in the Soviet Union, it is likely that they also concluded that effective BMD was likely to founder on financial and technical grounds. Deployment of systems on both sides was likely only to lead to increased deployment of ICBMs.

It was therefore in the interests of both the superpowers to prevent widescale deployment of ABM systems, and as a result the ABM treaty was negotiated as part of the strategic arms limitation talks (SALT) in 1972. The SALT I ABM treaty prohibited each country from deploying ABM defence systems at more than two sites. The 1974 protocol reduced this limit to one site on each side. In the event, the United States has no system deployed, and the Soviet Union has a light ABM ring around Moscow, which consists of thirty-two Galosh and sixty-eight SH-04 missiles.

Ballistic missile defence foundered in the Sixties because the capability which the technology could provide was of dubious effectiveness, either in protecting cities or even in a limited aim of protecting missile sites. The cost of deployment would have been very high, and, given its limited effectiveness, the enemy could counter at much less cost by deploying more ICBMs or more warheads. The effect on the strategic balance, and on stability, was uncertain. In the Eighties, the debate was opened again, with both superpowers investing considerable effort in investigating possible BMD technologies. Can the new technology promise a strategy which enhances stability and can be afforded, or are we seeing history repeating itself? To seek the answer to this question, we can examine the well-documented US proposals for a ballistic missile defensive system.

THE HIGH FRONTIER

The main elements of a next generation BMD system are summarised in the High Frontier study.[2] A layered defensive system which seeks to destroy most hostile ICBMs in their early boost phase is proposed. Those remaining would be re-attacked during their midtrajectory phase. Those warheads penetrating the first two layers would be attacked during the terminal phase by ground-based ABM systems. Civil defence measures complement this defensive programme. The study claims that:

> We can deploy in space a purely defensive system of satellites using non-nuclear weapons which will deny any hostile power a rational option for attacking our current and future space vehicles or for delivering a militarily effective first strike, with its strategic ballistic missiles on our country or on the territory of our allies. Such a global ballistic missile defence system is well within our present technological capabilities and can be deployed in space in this decade at less cost than other options that might be available to us to redress the strategic balance.[3]

Not everyone shares this degree of confidence in such a global BMD deployment. In early 1983, President Reagan set up a commission to review the strategic modernisation programme of the United States. Among other topics, the Scowcroft Commission looked at the question of BMD. They found that:

> Applications of current technology offer no real promise of being able to defend the United States against massive nuclear attack in this century. An easier task is to provide ABM defence for fixed hardened targets, such as ICBM silos. However, even this will be a difficult feat if an attacker can use a large number of warheads against each defended target.[4]

To look at these contradictory assessments in more detail, the technological, economic, strategic and arms control implications must all be considered.

ABM TECHNOLOGIES

Since the development of MIRV for strategic missile systems, the problem with countering any ICBM attack has been significantly complicated. Considerable benefit would therefore be gained by an ABM system which destroyed the missile in its early boost phase, before the independent re-entry vehicles had separated. Attack during this phase brings two other advantages: boosters are far more visible than RVs for detection, and it is more difficult to provide decoys which look like boosters than it is to simulate RVs. The major disadvantage is that boost-phase intercept of the ICBMs must take place over enemy territory. This leads to the requirement for a space-based target acquisition and interception system. If the critical boost-phase interception can be achieved from space, then the midtrajectory interception can also be space based. The terminal phase of the layered defensive system would be met by ground-based systems defending the expected target area.

A number of proposals have been made for possibly useful lines of research for the space-based systems. These can be divided into approaches based on directed-energy techniques and those dependent on kinetic energy kill mechanisms. The interception problem is formidable, with perhaps 2,000 boosters to be engaged within the first three minutes of their flight. The near-instantaneous response of directed-energy weapons, either lasers or particle beams, makes them an attractive option for this time-urgent task. Laser weapons would kill missiles either by burning through the booster skin, or producing a shock wave which caused mechanical damage. One calculation of the energy requirement has suggested that a 100 megawatt laser, using a mirror some four metres in diameter, could cause sufficient damage to a target in approximately one second.[5] If such a hydrogen chloride laser could be deployed in sufficient numbers it would require 1,000 shuttle flights for energy stores alone. This suggests that, for laser-based systems, a nuclear-pumped X-ray laser would need to be developed.

Particle-beam weapons seek to kill missiles through damage to their electronics. A charged-particle beam would suffer from dispersion from unpredictable fluctuations in the earth's magnetic field, and also from ionisation in the earth's atmosphere. A neutral-particle beam cannot be accelerated or focused. Any space-based particle-beam system would therefore need to accelerate a charged-particle beam, focus it and then neutralise it by stripping extra electrons from the atoms. This is a complex process, requiring technology which is currently beyond the state of the art. Consideration of the fuel requirements leads to similar conclusions to the laser, and the likely necessity of a nuclear explosive system for any workable particle-beam weapon. One analysis concludes 'all these difficulties point to the conclusion that even though it might be possible to construct an accelerator that would produce a beam of hydrogen atoms intense and energetic enough to cause some damage to the ICBM in its boost phase, a satellite-based neutral-particle-beam ABM system would not be practical.'[6]

Kinetic energy kill systems can either be of the rocket-propelled missile type, or of the hypervelocity gun system. The High Frontier study sees the deployment of carrier satellites with conventional missiles as a practical proposition. The total mass of such a system precludes comprehensive deployment. A more promising area for research into kinetic energy kill systems is to be the hypervelocity electromagnetic railgun. This system uses a high energy electromagnetic field to accelerate a hit-to-kill projectile to speeds of the order of 5–25 km per second. Again, the power requirements for such a system are prodigious, and are likely to need nuclear generators.

Whatever system is used to kill the ICBMs, it must be able to cope with a massive simultaneous attack. Target acquisition, tracking and pointing all give significant technological problems. Numerous counters to each weapon system are already in prospect. The vulnerability of the weapon system itself is a major cause for concern. Already antisatellite weapon systems are well advanced, and one analyst has suggested that space-based ABM systems would be particularly vulnerable to space mines.[7] In summary, the outstanding technical problems are the provision of adequate target acquisition, tracking and pointing systems, effective kill mechanisms at attainable energy costs, and protection of the weapon system itself. The Scowcroft Commission concluded that: 'At this time, however, the Commission believes that no ABM technologies appear to combine practicality, survivability, low cost and technical effectiveness sufficiently to justify proceeding beyond the stage of technology development.'[8]

Certainly the technology for terminal interception of ballistic missiles has improved since the sixties. A favoured option for this point defence of US ICBM sites is the Swarmjet. The system uses a launcher with 500 to 1,000 unguided rockets fired simultaneously. Between 5,000 and 10,000 rockets are assessed as giving an 85 per cent probability of intercepting a single re-entry vehicle within the lethal area of the Swarmjet system. Multiplying this requirement across all the possible targets with the more than 7,000 Soviet RVs currently deployed emphasises the importance of the boost-phase interception.

Technology points to the possibility of such a comprehensive layered defensive system, given infinite resources. The question is what can be achieved with finite resources, and whether it is worth doing.

THE ECONOMICS OF BMD

Given the uncertainties of the technologies discussed above, any assessment of the cost of deploying a full-layered BMD system must be speculative. The High Frontier study makes a detailed assessment of the costs of deployment. It states that the whole system could be deployed by 1990 for a total cost of $35 billion. However, 'cost data validity for the illustrative High Frontier programs range from hard to very soft.'[9] A comparison with the cost of the UK Trident programme, a much more modest affair, suggests that this may be an order of magnitude underassessed. The Defensive Technologies Study Team in the United States has already earmarked $11.4 billion merely for research to select appropriate technology paths for ballistic defence through either directed energy weapons or conventional interceptors.[10] The North Atlantic Assembly scientific and technical committee has assessed the cost for a total system in the range of $100–500 billion.[11] For a comparison the 1983 defence budget for the United States was a total of $240 billion.

Economic considerations are therefore going to play a major part in any decisions on the deployment of a BMD system. Given the economic difficulties that all nations are having in maintaining their force levels, it may be the overriding factor.

STRATEGIC IMPLICATIONS

Even if it is assumed that all the technological problems can be overcome, at a price which is affordable, what are the implications for the security of the West? None of the proponents of BMD make the claim that it can ever produce a 100 per cent effective defence. The exact percentage of 'leakage' is a matter of some debate. At the upper end of effectiveness, a full deployment would protect adequately against all contingencies except a massive retaliatory counter-city attack. At the lower end of the assessment, the system would merely reduce the vulnerability of the fixed land-based ICBMs from a pre-emptive first strike. Any assessment of the effect of BMD deployment on stability must depend on the prior assessment of effectiveness.

If a high 'leakage' rate, say greater than 10 per cent of incoming missiles, were found to be the likely value, then deployment could not be cost effective. The enemy could always counter more cheaply by increasing his missile numbers. The vulnerability would only be that of land-based systems, and the continuing invulnerability of submarine-launched ballistic missiles would ensure that deterrence continued on the same basis as before.

If the BMD deployment appeared to hold in prospect a real change in the strategic balance, what would the effect be then? A simultaneous deployment by both sides of an effective BMD system might make the possibility of fighting a conventional war both more likely and more acceptable, in that the risks of unacceptable nuclear destruction of the homeland had been reduced. Should one superpower consider that the other might achieve deployment first, then the prospects for stability could be of concern. The nation losing the race for deployment might feel so threatened that it needed to take pre-emptive action against the BMD system. Certainly, both sides will feel the need to investigate the technologies, to have the capability to deploy counter-measures, should the other side produce an operational BMD system.

In summary, because perfect defence is unattainable, nuclear deterrence, through the threat of massive retaliation, should remain effective. Depending on the level of BMD effectiveness, it might be that the prospect of limited war between the superpowers became possible again. In any event both would need to deploy appropriate counter-measures, which could include significant increases in their nuclear systems.

ARMS CONTROL IMPLICATIONS

Research and development on ABM systems is permitted by the ABM Treaty. The deployment of anything beyond a single-point defence ground-based system would certainly require a renegotiation of the Treaty, and in all probability termination of it. In the near term, research can continue, but at the stage where field testing of new systems was necessary, the provisions of Article V and the Treaty would have to be broken. At this stage it would seem unlikely that either side would feel it had much to gain from a new, and less restrictive, treaty. It is, of course, possible that the demise of the ABM Treaty would signal the end of the other arms limitation protocols with a view to expanding the offensive capability on both sides.

The Case for BMD

For the critics who express concern about technical difficulties, shortsightedness of eminent scientists in the past is cited. Caspar Weinberger, addressing the National Press Club in May 1984 reminded his audience of the Chief of Staff during President Truman's incumbency in 1945, who stated in regard to the atom bomb that: '*The bomb will never go off, and I speak as an expert in explosives.*' He went on to cite the statement of Dr Vannevar Bush, the President of MIT in 1945, that: '*The whole idea that you could launch a missile from one continent and hit a target 4,500 miles away is too preposterous and absurd even to contemplate.*'[12] The strategists who worry about the effects of the deployment of strategic defence systems on stability and on the prospects for arms control are reminded that the Soviets have an advanced research programme in this field. Nor are the strategic arguments accepted by the proponents of SDI:

Not only can the Strategic Defense Initiative strengthen deterrence it can also enhance the opportunity for arms reductions. Strategic Defense can strengthen deterrence by making it highly uncertain or impossible, for a missile attack to achieve any rational military objective. And it can enhance arms control by devaluing nuclear ballistic missiles, which will create powerful incentives for sharp reductions in their numbers—reductions that would enhance the security of the United States, its allies, and the Soviet Union.[13]

For those who worry about the costs of such a deployment, regardless of the strategic or technical problems, it is argued that the costs for such a system will be affordable, and indeed the strategic defence will be a cost-effective way of pursuing deterrence.

I think we must approach the problem that the President posed with the objective of making it affordable. I think we can. You also have to remember that we spend approximately 15% of our annual defense budget on strategic forces, and strategic force modernisation. And that is a whopping sum of money.[14]

The debate then is becoming more heated, and more facts are emerging into the public domain. Has greater public awareness shed any more light on what is either possible or desirable?

The Strategic Defense Initiative

The first question to be answered is the nature of the Strategic Defense Initiative. President Reagan appeared to be offering the prospect of a magic bubble of invulnerability which could cover America and Europe when he said: '*What if . . . we could intercept and destroy ballistic missiles before they reached our own soil or that of our allies?*'[15]

This reflected the earlier strong assertions of the High Frontier study group, who believed that a non-nuclear defence of cities was an early possibility. Following the Presidential directive, the SDI has been formalised and, in a statement to the Senate Committee on Armed Services at the end of 1983, the Department of Defense explained the proposals.[16]

It must be understood that our programme is not a system development program. No decision has been made to develop and deploy any weapons or other elements of the potential system. Our state of knowledge of the relevant technologies is inadequate. Consequently, the aim of this program is to improve our knowledge of the relevant technologies by providing firm calculations and experimental evidence on what such technologies could do and at what cost. But in other words, our program is a research effort to provide the evidentiary basis for an informed decision on whether and how to proceed into system development. For a full, multi-tiered system, we expect to complete the provision of this evidentiary basis by the early 1990s. A careful analysis of defensive strategy and concepts of operation is an essential element of this basis for an informed decision. Our program also includes such an analysis.

The statement went on to announce that a separate Strategic Defense Technology Office was being established. The research programme would look at five different areas:

surveillance, acquisition and tracking
directed-energy weapons
kinetic energy weapons
systems analysis and battle management
support programmes

The Technologies. The aim of the SDI research programme into surveillance, acquisition and tracking is to develop the capabilities needed to identify targets during all phases of a ballistic missile flight path. These complex technologies must be such that they can operate reliably even under nuclear attack conditions. The initial element of the programme is concerned with detecting optical, infrared and radar signatures of re-entry vehicles. Optical imaging using lasers rather than radar beams is being investigated, and a substantial amount of effort is being devoted to developing infrared sensors with near realtime signal processing. The first demonstrated systems are expected to be a boost-phase detection and tracking system, a midcourse long wavelength infrared sensor for tracking and discriminating objects during midcourse, and terminal detection systems using both optical and radar trackers. This element of the research programme is seen as the crucial determinant of whether defensive systems of the necessary capability can ultimately be built.

Directed-Energy Weapons. The directed-energy weapon research programme is looking into space-based lasers, ground-based lasers, space-based particle beams and nuclear-driven directed-energy weapons. Research must consider the type of beam generators, how to control the beams, and will look into the acquisition, tracking and pointing problems, including the provision of large optics for directing the beams. The DOD expects to be able to demonstrate by the end of the decade '*a capability to control wavefront errors, maintain beam alignment within the system, compensate for atmospheric affects; and provide the components necessary to transmit and control the high intensity beams.*' The work on large optics is not expected to come to fruition before the 1990s. Again, once all the research on the component parts has been carried out, an assessment can be made on whether it is possible to move on to the more complex complete integrated system.

Kinetic Energy Weapons. Less exotic is the programme for development of appropriate kinetic energy weapons for destroying missiles. These would include high performance interceptor missiles, high velocity gun systems and the associated fire and guidance control and warhead design. The practicality of a kinetic energy kill solution to ballistic missile defence was demonstrated by the US Army Homing Overlay Experiment (HOE) of 10 June 1984.[17] In this demonstration, an incoming ballistic missile was destroyed by an interceptor which had been boosted out of the earth's atmosphere by a Minuteman first and second stage. Once out of the atmosphere, the homing stage intercepted the incoming warhead by acquiring and tracking it with its infrared sensor. A 4.5-metre metal umbrella was unfurled, and the kinetic energy of the collision was so great that no explosive charge was necessary to destroy the warhead. *This single successful test* *(the first three had failed) cost over $300 million*, which indicates the order of magnitude of research costs in each of these high technology areas.

System Analysis and Battlefield Management. The aim of the system analysis and battle management programme is to develop the technologies which will eventually allow the system to be rapidly responsive, completely reliable and survivable while still being controllable. Within this area the research will look into the nature of the threat, concept of operations, system design, and the overall technology assessments for each level of the multitiered defensive system. Much of the research will be into high performance processors and software, and the associated communications network. The evaluation of the system will depend a great deal on simulation, and development of effective simulation models will also be necessary.

Support Programmes. A major difficulty with the establishing of a comprehensive defensive capability, is the scale of the logistic support required. As a result, the support programme's element will look into the energy requirements for the various systems, and the associated logistic bill, and how it may be met.

The Aim of SDI. The evidence to the Senate Committee concluded by saying that:

> In summary, we believe that an effective defense against ballistic missiles could have far-reaching implications for enhanced deterrence, greater stability, and improved opportunities for arms control. Our efforts do not seek to replace proven policies for maintaining the peace, but rather to strengthen their effectiveness in the face of a growing Soviet threat. The essential objective of the Strategic Defense Initiative is to provide future options to diminish the risk of nuclear destruction and to increase overall stability. By the end of the decade we will have conducted a number of ground, airborne, and space experiments. The knowledge gained from these tests will help to identify those technologies that are most promising and support decisions in the early 1990s on whether and how to proceed with development of ballistic missile defenses. The planned funding for the various areas of the Strategic Defense Initiative are shown in Table 5.1 rising from $1 billion in 1984 to $7.4 billion in 1989.[18]

Table 5.1: *Planned DOD funding for the strategic defence initiative (in millions of dollars)*

	FY84	FY85	FY86	FY87	FY88	FY89
Surveillance, acquisition and tracking	367	721	1,491	1,944	2,656	3,331
Directed energy	323	489	1,020	1,222	1,377	1,683
Kinetic energy	196	356	870	1,274	1,514	1,437
Systems analysis and battle mgmt.	83	99	138	227	260	288
Support programmes	24	112	271	322	453	667
TOTAL	993	1,777	3,790	4,989	6,260	7,406

The Critics' View

The critics of SDI argue that these vast sums need not be spent to carry out research into what is already self evident. They argue that the development of such antimissile technologies is irrelevant and cripplingly expensive at best, and strategically dangerous at worst. By postulating various levels of success in the technological development, it is possible to speculate on the effects on stability and deterrence.

The SDI proposals are exclusively designed to research into counters for ballistic missiles. Nuclear threat is not exclusively from ballistic missiles. Low altitude air-breathing systems, such as manned aircraft and Cruise missiles, are already deployed. The use of nuclear suitcase weapons, surface ship bombs, and even depressed trajectory ballistic missiles, would all continue to be possible. Indeed it is likely that deployment of such systems, which are less easily verified for arms control purposes, would accelerate in the light of any success in ballistic missile defence. The options for such delivery systems mean that ballistic missile defence could not even provide security against small nuclear powers.

Currently both the US and the Soviet Union have around 10,000 strategic warheads each.[19] If 1 per cent of those warheads exploded, it would be sufficient to destroy the nation as a twentieth-century power. The critics would argue that it will also be cheaper to increase the number of warheads, to ensure that sufficient could get through.

While the offence has the advantage in cost today, it is likely that this advantage could be multiplied by a number of already available counter-measures. These include decoys and penetration aids to the missiles, reducing the boost-phase time for missiles, the use of space mines, ablative surfaces on missiles, rotating missiles to absorb directed energy and increase the power requirement. The efficacy of these measures might be great or small: the difficulty would be that neither side would know until deterrence failed.

Looking at the effect of a deployment of a ballistic missile defence system on stability, a number of concerns have emerged. Some have suggested that the Soviet Union would be tempted to attack before deployment, fearing that it would be losing its nuclear capability after an effective ballistic missile defence was in place. It is also possible that the Soviets could perceive such a deployment as a prelude to a Western pre-emptive first-strike strategy. In either event, such perceptions would be destabilising.

The Way Forward

The SDI is not going to provide the realisation of the dream of a world secure from nuclear threats. The research programme, for that is what it is, being undertaken will investigate a number of very interesting military specialisations. In the future, there will be an increasing need to be able to counter conventional ballistic missiles, and much of the work done on the SDI will pay off in this respect. The sensor technologies will have a great benefit for the weapons of tomorrow. Directed-energy systems may have a greater role to play on the battlefield of tomorrow. Certainly improvements in command control and communications will benefit all aspects of the military organisation. The impetus which President Reagan's SDI has given means that much useful military research is being undertaken. In any event, this research is also important to ensure that appropriate counter-measures are available should the Soviets deploy some form of ballistic missile defensive system.

It is what comes beyond SDI which should be of concern. The diversion of a large portion of US defence funds into a system of limited capability and dubious strategic impact is not a good prospect for Western security. It is to be hoped that the research which both the United States and the Soviet Union are currently undertaking into ballistic missile defence will again convince each side of the benefits of arms control in this field.

NOTES

1. Statement Before Committee of Armed Services (24 February 1965).
2. Daniel O. Graham, *The Non-Nuclear Defense of Cities* (Cambridge, 1983).
3. *ibid.* p. 14.
4. 'Report of the President's Commission on Strategic Forces' (6 April 1983) p. 9.
5. 'Laser weapons', *Scientific American* (December 1981) p. 39.
6. 'Particle-beam weapons', *Scientific American* (April 1979) p. 44.
7. R. L. Garwin, 'Ballistic missile defence', (unpublished) (9 June 1983).
8. 'Report of the President's Commission on Strategic Forces' (6 April 1983) p. 12.
9. Daniel O. Graham, *The Non Nuclear Defense of Cities* (Cambridge, 1983) p. 59.
10. *Aviation Week and Space Technology* (24 October 1983) p. 50.
11. 'North Atlantic Assembly Report on Ballistic Missile Defence' AA205 STC(83)8 Para 27.
12. Secretary of Defense's address to the National Press Club is quoted in *National Defense* (September 1984) p. 71.
13. Caspar Weinberger, Secretary of Defense, speaking in California on 24 October 1984.
14. George A. Keyworth II, Science Adviser to the President, in an interview with the *Washington Times* on 19 October 1984.
15. President Reagan, Speech to the Nation, 23 March 1983.
16. Statement of Richard D. de Lauer, Under Secretary of Defense for Research and Engineering, before the Special Committee on Strategic and Theater Nuclear Forces, Committee on Armed Services of the United States Senate, 13 December 1983.
17. For full description *see Flight International* (15 December 1984) pp. 1642–3.
18. M. S. Smith, 'Star wars: anti satellites and space based BMD', Congressional Research Services Issue No 1B81123 of 18 July 1984.
19. US/Soviet Military Balance by John Collins, August 1984, gives figures of 10,588 US strategic warheads to 9,181 warheads.

6 US Offensive Capabilities and Nuclear Strategy

Steve Smith

Two events in early 1986, the destruction of the Space Shuttle launcher Challenger on 28 January and the explosion shortly after the launch of a Titan 34D rocket booster on 18 April, provide vivid examples of the centrality of space-based systems in the US strategic posture. Ever since the dawn of the nuclear age, analyses of the US–Soviet strategic balance have concentrated almost exclusively on static indicators such as the number of ICBMs and SLBMs and the capabilities of the warheads on them. Yet in the past few years attention has shifted to the issue of the command and control of these systems, and it is in this area that space-based assets play a vital role. This chapter is concerned with the US offensive capabilities in space and will concentrate on the implication of developments in this area for US nuclear strategy.

Before turning to examine the role of space-based assets in US nuclear strategy it is essential to make one qualification to the terminology used here: clearly there is no simple, uncontentious distinction that can be made between offensive and defensive systems, if only because each side will classify any given system of the other side in a way that is different to the way it is seen in that other country. To give only the most obvious example, ASAT systems are seen as defensive by the state that possesses them and as offensive by the other superpower. This is because each of the superpowers sees the role for their ASAT as preventing the other superpower from co-ordinating a pre-emptive attack or as crippling their C^3I facilities during any protracted nuclear conflict. Yet, of course, from the perspective of the other superpower, the same ASAT system must be seen as a force multiplier, the use of which would make it easier for that state to carry out a disarming pre-emptive attack or to gain advantage in a protracted conflict. We cannot hope to reach some neutral technical judgement as to which systems are offensive or defensive since this classification is axiomatically dependent upon perceptions. The approach adopted in this chapter is that we will look at those systems that are not capable of attacking targets on the ground and those that are not capable of being used to attack ballistic missiles in flight. In short, we are concerned with those space-based assets that fall between ballistic systems that travel through space and the candidate components for any SDI architecture.[1] This means that we will focus on the US ASAT system and on those satellites in space that serve as the eyes and ears of the offensive component of the US arsenal.

The impact of the loss of the Challenger shuttle and the Titan 34D booster on US offensive capabilities is that these launch vehicles are central to the operation of US C^3I systems in space. The Titan loss in April 1986 was the second successive failure of that system in eight months; in August 1985 a Titan 34D booster exploded shortly after launch. Although many press reports indicated that on both occasions the vehicle was carrying a KH-11 reconnaissance satellite, it was leaked in late April 1986 that the 18 April 1986 launch in fact carried a 'Big-Bird' KH-9 satellite.[2] The crucial fact is that with both the Shuttle and the Titan grounded pending investigation of the causes of the failures, the US has no capability for launching heavy payloads into space. This means that there is no way of placing into either polar orbit or lower inclination and geosynchronous orbits any heavy satellite. The damage to the launch area at Vandenberg (where the April 1986 Titan launch took place) means that even were a system available there could be no location for launches into polar orbits until that area is repaired. The importance of the absence of this capability is that the US has only one KH-11 reconaissance satellite in orbit; this was launched on 4 December 1984 and is halfway through its expected lifetime. Usually the US maintains two such satellites in orbit, so there will be a period of a few months during which there will be limited satellite coverage, and, of course, there would be a considerable problem were the KH-11 in orbit to malfunction. The KH-11 was scheduled to be replaced by the KH-12 in late 1987, but this is so large that it can only be launched by the Space Shuttle. The Big Bird (KH-9) is a satellite designed to provide high resolution photographic information, which is returned to earth by film capsules. It has a lifetime of about 300 days, is launched into a sun-synchronous orbit with an inclination of 96.4 degrees and has an apogee and perigree of approximately 155 and 100 miles respectively.[3] The KH-11 is crucially different to the KH-9, since it transmits its information in real time, and uses a digital radio pulse system to provide imaging data. It, too, is launched into a polar sun-synchronous orbit, inclined at 97 degrees, but travels in a more elliptical orbit, with an apogee and perigree of 320 and 160 miles; it is also manoeuvrable so as to be able to divert to observe any areas of particular interest, and has a longer lifetime, of about four years.[4] The KH-12 replacement of these satellites causes a problem, since there are reportedly only a few KH-9 and KH-11 satellites left in the stockpile, and the KH-12 will have to await the report on the Challenger accident and any modifications to the boosters that result. A final problem is that the DSP early-warning satellites, launched from Cape Canaveral into geosynchronous orbits, also use the Titan 34D. None was launched in 1985, and the fact that two or three have been launched each year indicates that 1986 and 1987 would have seen several such launches. The Titan and Challenger failures therefore look like having serious implications for the US's ability to gain photographic information on Soviet strategic deployments and for the reliability of the US early-warning system. There are alternative systems, such as the SR-71 reconnaissance aircraft, but the extent of the reliance on the Titan and Space Shuttle to launch these components of the US C^3I system is considerable.

This brief excursion into the problems currently facing the US is important in directing our attention to the linkage between space-based assets and offensive capability. All too often the crucial variable is defined as the balance of offensive forces *per se*. Complex calculations about the vulnerability of either side's ICBM force are undertaken, with improved accuracies on either side being treated as a major cause for concern.[5] Yet, three recent studies by Bruce Blair, Paul Bracken and Daniel Ford[6] have argued that such a focus is very misleading. These three studies have looked at the vulnerability of the US C^3I system, and all conclude that a massive vulnerability has existed in this system for the last twenty years. In the most detailed of the studies, Bruce Blair points out that a prudent Soviet planner would not need to worry about destroying US ICBMs; a far easier task would be to disable them by attacking US C^3I facilities. In short, these three studies indicate that a central aspect of the offensive military balance is the vulnerability of their C^3I systems, without which offensive forces could not operate. In this light, the Challenger and Titan failures are most salient examples of the reliance of the US on space-based assets in the operation of their offensive forces.

It is because of this reliance that space was 'militarised' long before President Reagan's 'Star Wars' speech of 23 March 1983. Given the central role of early-warning, reconnaissance and communications satellites for the operation of US strategic forces, much of the material in the literature on US–Soviet strategic systems has been focused on a rather narrow area of that competition. After all, it is no good having very accurate ICBM warheads if they cannot be launched before being destroyed, because of their vulnerability to an attack on their C^3I linkages. It is for precisely this reason that both the US and the USSR have been involved in developing systems to attack these C^3I assets. This chapter will now turn to examine the US attempt to develop an ASAT system before discussing its implications for US nuclear strategy. Because much of this information has already been covered elsewhere in this volume we will limit the survey to a brief overview of the history of the US ASAT.[7]

The US began work on an ASAT system in 1959 when a programme known as Bold Orion was begun, although this was essentially an adaptation of a programme to develop air-launched ballistic missiles.[8] During the 1960s there were a number of projects, known as Programs 437, 706, 922 and 505. Of these only Programs 505 and 437 led to deployment, and, contrary to conventional wisdom, the US was the first to deploy an ASAT. This system (Program 505) used a Nike-Zeus missile in an ASAT role and was declared operational in August 1963. The potential of Program 437, which used a Thor missile in an ASAT role, led to the phasing out of Program 505 in 1967. From 1964 to 1970 16 firings of the Thor (Program 437) ASAT were carried out, although the Limited Test Ban Treaty of 1963 prevented any detonation of the nuclear warhead that would have destroyed the target. During the late 1960s the disappearance of the threat that Programs 505 and 437 were intended to counter—orbital bombs (FOBs)—meant that interest in the system declined. In 1970 it was decided that, because of its cost and the collateral effects of using a nuclear detonation in space (for example, on US satellites!), such a system was not worth proceeding with, although it remained nominally operational until 1975.

The current US ASAT project was begun in the early 1970s with full scale development commencing in 1977. The system involves the 'direct ascent' of an interceptor to its target, in contrast to the Soviet co-orbital ASAT system tested between 1968 and 1982.[9] The interceptor consists of an MHV on a two-stage missile. The system is mounted under an F-15 fighter, which takes the missile up to a height of about 10 kilometres; the SRAM first stage and the Altair second stage take the interceptor up another 500 kilometres, from where the MHV uses its eight heat-seeking infrared sensors to acquire the target, and then fires small rocket thrusters to ram the target. Destruction is achieved by the velocity of impact. The first flight test occurred on 21 January 1984, when an ASAT booster, without MHV aboard, was launched against a point in space. A second test in November 1984 was targeted against an infrared emitting body (a star) to test the ability of the MHV to distinguish between its target and the background infrared emission of space. Both of these were considered to be successful. The third, and most important, test took place on 13 September 1985; in this test, the complete system was launched against a target satellite. The MHV successfully intercepted the target.[10] A further nine tests of the system are planned.

However, the US ASAT programme has run into deep trouble in Congress. To understand this, one has to go back to the late 1970s; under the Carter Administration a series of negotiations took place with the Soviets to try and reach agreement on an ASAT ban.[11] This was known as the two-track policy: under PD-37 the Administration pledged itself to try and achieve an ASAT ban, but to continue research on ASAT in case this did not transpire. The talks took place between June 1978 and June 1979, but no agreement was reached. The Soviets wanted to obtain a guarantee that the Space Shuttle would not be used as an ASAT weapon, but would not agree to dismantling their own ASAT system. They accepted the idea of a non-use agreement, but argued that this would only apply to US and Soviet satellites (i.e. it omitted NATO and Chinese systems). The US position differed crucially in that it wanted a non-use agreement coupled to a testing moratorium, but insisted that the Soviets dismantle their own system. If they would not do that, then the US would need to develop their own ASAT. The talks were finally stymied by the problems SALT 2 was facing in late 1970, and then by the Soviet intervention in Afghanistan. In 1981 the Reagan Administration quickly indicated its opposition to an ASAT agreement by refusing to discuss a Soviet draft ASAT treaty and refusing to offer any counter-proposal. The desirability of developing a US ASAT capability was explicitly mentioned in the Reagan 'Strategic Modernisation Program' announced in October 1981.

In the summer of 1983, Andropov announced that the Soviets would not test their ASAT until such time as the US tested theirs. Whilst one can argue that the Soviets had already conducted some twenty tests, and therefore did not need to test further, it is salient to note that all their tests with an infrared heat-seeking sensor had failed. This left them with a radar-homing sensor as the only tried and tested target acquisition method available, and this form of acquisition is relatively easy to jam. The severe operational limitations on their system also made the offer less one-sided than might at first appear to be the case.[12] One can debate how near to agreement the two sides had been in 1978–9, and it is possible to envisage ways of overcoming the differences between the two sides, but the central point is that the Reagan Administration treated ASAT in exactly the same way as it treated strategic nuclear forces generally: to get a worthwhile agreement the US had to be in a position of superiority. There was not the political will to reach agreement, and attempts to do so would have to await the completion of the testing of the US ASAT.

But the Soviet moratorium had a significant effect in Congress. Congress had been calling for a resumption of ASAT negotiations since 1981, and the Soviet moratorium indicated a willingness on their part to discuss an agreement, as did the two Soviet draft treaties on an ASAT ban presented in 1981 and 1983. On 18 July 1983 Senator Paul Tsongas was successful in getting an amendment made to the FY 1984 DoD Authorisation Act. This denied the use of funds for any testing of the US ASAT unless the President could state that the US was negotiating in good faith with the Soviets and that such tests were necessary for US national security. This was subsequently watered down after considerable lobbying from the White House; the new restriction required the President to certify that testing was necessary to avert damage to US national security and would not gravely impair the prospects for negotiation. Accordingly, President Reagan certified shortly before the 13 September 1985 test of the complete US ASAT system that his Administration had been 'endeavoring in good faith' to negotiate the 'strictest possible limitations' on ASAT weapons.[13] This was seen by many in Congress as a mockery of Congressional restrictions, since no negotiations had taken place. The Administration's response was that it had offered to discuss an ASAT ban with the Soviets but only as part of a wider agreement on strategic and intermediate nuclear forces.

It is absolutely clear that the Administration had no interest whatsoever in an ASAT ban, and was therefore linking any discussions to success in other arms control arenas. It is also clear that the disdain the Administration showed for Congress was crucial in the decision taken on 16 December 1985 to halt any funding for further ASAT tests against targets in space unless the Soviets ended their moratorium.[14] Edward Aldridge Jr, Air Force Under-Secretary, stated that he was 'disappointed with the congressional action . . . which in effect puts the fate of our testing of this critical defense program in the hands of the Soviets'.[15] Another view was that of Representative Les Aucoin, one of the major supporters of the ASAT test ban, who stated that Congress had done 'more for arms control in five hours than has been done in the last five years'.[16] Until such time, then, that the Soviets end their moratorium, or until Congress reverses its policy, the US ASAT programme is limited to tests against points in space, as distinct from target objects. Having offered a brief overview of the history of the US ASAT we need to focus on two related issues: *firstly* what does the development of the US ASAT tell us about the direction and nature of US nuclear strategy? *Secondly*, what would be the strategic implications of the continued development of ASAT systems by both superpowers? The answers to these questions will allow us to discuss the desirability of an ASAT arms control agreement.

To understand the implications of the development of the US ASAT for US nuclear strategy we need to see that development in the light of the direction of Administration thinking on strategy over the last decade or so. The first point to make is that there is a very popular myth concerning US nuclear strategy, one which sees the US accepting a policy of MAD from the mid 1960s, a policy that involved the targeting of cities, only for this to be replaced with a war-fighting posture, involving military targets, in the late 1970s. This is dangerously misleading. As many recent accounts of the history of US strategic targeting have indicated,[17] the US has never solely targeted Soviet cities. The confusion has occurred because of the similarity of two key terms—assured destruction and mutual assured destruction (MAD). The latter refers to a statement of fact, that the result of nuclear conflict was seen in the 1960s and 1970s as being mutual assured destruction. The former was the minimum targeting capability defined as necessary by Secretary of Defense Robert McNamara after his famous speech at Ann Arbor in February 1982. In that speech, McNamara argued that US strategy should move away from targeting population centres towards the targeting of military forces—the response should be flexible. It was that shift that required a statement of what would be the minimum level of destruction to be inflicted on the Soviets, because the absence of such a minimum would have allowed the Air Force to justify massive expenditures on a range of strategic systems deemed by them necessary for flexible response. As McNamara said in a memo to the President in November 1962: 'It has become clear to me that the Air Force proposals . . . are based on the objective of achieving a first-strike capability'.[18] Thus, McNamara publicly announced that the US needed to be able to destroy 25–35 per cent of the Soviet population and 67 per cent of industrial capacity in order to ensure deterrence. The paradox is that this was taken as being US weapons *employment* policy; it was, in fact, only *declaratory* policy. There was, then, no sense in which US targeting in the 1960s was directed primarily against cities, although this was evidently the accepted public wisdom (and also a common view amongst strategists!).

It was because of the popularity of the notion that US systems were targeted against civilian population centres that the changes in declaratory policy made in the 1970s and 1980s were seen as shifting the US to a war-fighting posture. In reality the changes were far less dramatic in terms of their impact on US weapons *employment* policy. The first public indication of a 'shift' in US strategic policy came in 1974 when Defense Secretary James Schlesinger revealed that President Nixon had signed a new Nuclear Weapons Employment Policy under National Security Decision Memorandum (NSDM) 242. Under this directive, a set of targeting sets was outlined, concentrating on military targets such as missile silos. Population centres were not to be targeted *per se*, except as a last resort, when the objective was to be to destroy 70 per cent of the Soviet economic recovery base, which translates into attacks on about 200 Soviet cities, and thereby involved some one-third of the Soviet population.[19] The thrust of NSDM 242 was threefold: to concentrate retaliatory capability on military targets; to control the escalating process; and to withhold certain types of targets, such as population centres. These changes were incorporated into SIOP-5 in 1976.

In the Carter Administration there was, under PD-18, a review of nuclear-targeting policy, which resulted in three further Presidential Directives, PD-53, PD-58 and PD-59.[20] PD-53, signed on 15 November 1979, called for improvements in the C^3I area, so as to allow for 'connectivity between the National Command authority and strategic and other appropriate forces to support flexible execution of retaliatory strikes during and after an enemy attack'.[21] PD-58, signed on

30 June 1980, concerned the maintenance of the continuity of government in the face of a nuclear attack. PD-59, leaked in August 1980 and officially entitled 'Nuclear Weapons Employment Policy', changed US strategy in two main ways. The most significant of these was to downgrade the emphasis accorded to economic targets under NSDM-242 and emphasise more military and political targets. The former already accounted for over half the 40,000 targets in SIOP-5; the latter were also already targeted, but PD-59 explicitly emphasised the importance of targeting the political leadership. The second change concerned PD-59's requirement that the US be able to fight a *protracted* nuclear war: this has crucial implications for US reliance on C^3I facilities. But PD-59 was essentially evolutionary rather than revolutionary. The changes made were to the emphasis placed on various sets of targets, and the options required by PD-59 were all concerned with retaliatory strikes. This strategy became known as the countervailing strategy.[22]

Under the Reagan Administration a further move away from the minimum requirements of assured destruction occurred. NSDD-13, signed in October 1981, re-affirmed the principles underlying NSDM-242 and PD-59 and increased the emphasis on nuclear options during a protracted war. According to the Defense Guidance Plan 1984–8, leaked to the press in May 1982, this emphasis was required in order to allow the US to prevail in a protracted conflict. The United States must be in a position 'to force the Soviet Union to seek earliest termination of hostilities on terms favorable to the United States'.[23] The main requirements for achieving such an outcome, according to the leaked reports, were: 'forces capable . . . of attacking a wide range of targets, even when retaliating to a massive strike received without strategic warning. . . . Employment plans that assure US strategic nuclear forces can render ineffective the total Soviet . . . military power structure through attacks on political/military leadership and associated control facilities, nuclear and conventional forces and industry critical to military power. . . . Forces that will maintain, throughout a protracted conflict period and afterward, the capability to inflict very high levels of damage against the industrial/economic base of the Soviet Union and her allies so that they have a strong incentive to seek conflict termination short of an all-out attack on our cities and associated assets . . . [and C³I systems to] provide the capability to execute ad hoc plans, even subsequent to repeated attacks.'[24] This is explicitly a war-fighting strategy, and includes phrases that are consistent with claims that the US is seeking a war-winning strategy.[25] But, crucially for our purposes, what NSDD-13 and the Defense Guidance Plan emphasise is the importance of communications during any protracted conflict, a concern evident in PD-53 and PD-58.

Whilst most commentators concentrated on the changes in declaratory policy represented by the directives we have discussed, such an emphasis is misleading in two ways. First, it assumes that population centres were the primary concern for SIOP until 1974, and this was not the case despite the public presentation of US strategy. The changes from NSDM-242 to NSDD-13 were indeed evolutionary and essentially publicly codified the reality of US targeting since the mid 1960s. The changes that were introduced in targeting options were towards greater flexibility. This is of central importance to the role of ASAT, since any increase in targeting flexibility axiomatically places more dependence on C³I facilities. More saliently, the drift towards war fighting and war winning under the Carter, and especially the Reagan, Administrations requires a capability to disrupt the Soviets' ability to target the US C³I and political authority structure. In short, if you want to prevail in a nuclear conflict you want to be able to prevent the Soviets from co-ordinating their targeting—you want to attack their C³I facilities, and this implies some ASAT capability.

The second sense in which a concentration on the strategic implications of NSDM-242, PD-59 and NSDD-13 is misleading is that the most important changes they led to were in the C³I field. The directives were not central in justifying strategic modernisation, nor did they fundamentally change the nature of US targeting. What they did do was to draw attention to the importance of C³I facilities and to the weaknesses in the then-existing US capabilities in this area. It is therefore not surprising to find that improvements in C³I have been the centrepiece of the Reagan Strategic Modernisation

Plan. As Defense Secretary Weinberger stated in his FY 1984 annual report, the Administration has given the 'highest priority to increasing the ability of our strategic force management systems not only to survive but to remain capable of performing their basic functions throughout a sustained sequence of Soviet attacks'.[26] Expenditure on strategic C^3I systems was expected to grow in real terms at an annual rate of 11.9 per cent from FY 1983 to FY 1986.[27]

Seen in this light, the attitude of the Reagan Administration to a ban on ASAT systems seems less puzzling. The key issue is that you see ASAT differently if you simple want assured destruction than if you want a war-fighting strategy. If your strategy is to ride out an attack and then retaliate against cities, then you have no strategic need to destroy the other side's space-based assets. Of course, you want to be able to launch your own retaliatory forces, and this has considerable implications for C^3I given the problems of communicating with submarines. But, if your strategy calls for fighting a protracted nuclear war (as envisaged under PD-59), and especially if you wish to prevail—as distinct from countervail—(as in NSDD-13 and the Defense Guidance Plan), then you will want to attack Soviet space-based assets. Doing so will disrupt their ability to fight a protracted nuclear war by making it difficult for them to acquire up-to-date targeting information. This is precisely why the Reagan Administration has been such a strong proponent of the US ASAT, and has been so firmly opposed to an ASAT ban.

This US commitment to developing an ASAT system, as has been pointed out, is currently curtailed by Congressional restrictions on the use of funds, but the testing of the US ASAT against points in space will continue, and so will the development of the system in general. We can now turn to discuss what would be the strategic implications of the possession of ASAT systems by both superpowers.[28] The first point to be made is that satellites are essential to notions of limited nuclear war. If neither superpower could have the data provided by satellites then it would be difficult to develop counter-force strategic-targeting options, and it would be impossible to be flexible in targeting during a nuclear conflict. This leads to a more general consideration, which is crucial to appreciating the impact of both superpowers possessing ASAT systems, and this concerns the role of satellites in nuclear strategy. Unless we understand what role satellites have in the strategies of the superpowers we will not be able to assess the potential effects of ASAT attacks on these strategies. The central issue is *not* one of the effectiveness of the US ASAT compared with the Soviet ASAT, but concerns the effect on US strategy of a Soviet ASAT capability compared with the effect of a US ASAT on Soviet strategy. Specifically, is the United States better off in an environment where there are no ASATs than in one where both sides possess them?

This requires a brief discussion of the current reliance on satellites by the US and the Soviet Union. Satellites perform a number of functions related to the command and control of nuclear forces. Of these the most important are: early warning, communications, reconnaissance, electronic intelligence gathering and navigation.[29] Each of these has clear implications that impact on the flexibility of each side's nuclear strategy. These satellites are placed in different orbits, and it is the height of the orbit that is the most important factor when assessing their vulnerability to the US and Soviet ASATs.[30] There are five main types of orbits:[31] (a) low orbits. These are orbits up to about 1000 km and the US and the USSR place their reconnaissance and electronic intelligence gathering satellites in these orbits. The Soviets also use these orbits for their meteorological and navigation satellites; the US does not. (b) 'Molniya' orbits. These orbits are highly elliptical, with apogees and perigrees of approximately 40,000 and 500 kms. Satellites following such orbits spend most time (about 8 out of their 12 hour orbital periods) over the northern hemisphere before coming down very low over the southern hemisphere. The Soviet Union bases its early warning satellites in Molniya orbits, as well as most of its communications satellites (the need of the Soviets to communicate with its bases in the Arctic means that Molniya orbits are necessary, since satellites in geosynchronous orbits, because of the curvature of the Earth, are not in line of sight with polar latitudes). (c) Semisynchronous orbits. These orbits are about 20,000 km above the earth, and satellites at that height orbit the earth twice a day. The US NAVSTAR navigation system is being placed in such orbits, as is the similar Soviet system GLONASS. (d) Geosynchronous orbits. These are located at a height of 36,000 km, with an inclination of O°. Satellites in this orbit circle the earth at the same rate as the earth rotates, and thus appear from the ground to stay over the same spot. The US places its communications, meteorological and early-warning satellites in this orbit, whereas at present the Soviets only place some of their communications satellites in geosynchronous orbit. (e) High orbits. The US places its nuclear detection (Vela) satellites at a height of about 110,000 km, but these are now being replaced by the Defense Support Program (DSP) early-warning satellites, placed in geosynchronous orbit; the secondary function of these is to detect nuclear explosions. The Soviets have no satellites in these high orbits.

This rather different pattern of usage of these orbits by the US and the Soviet Union leads to important asymmetries with regards to the threat posed by ASAT systems. The US ASAT has an operational height limit of about 500 kilometres; the Soviet ASAT is estimated to have a ceiling of 1500 kilometres. From the above survey of the usage of the various orbits we can see that the Soviet ASAT threatens only US reconnaissance and electronic intelligence gathering satellites, whereas the US ASAT threatens similar Soviet systems and their meteorological and navigation satellites. More saliently, the US ASAT, if it was based in the southern hemisphere, could threaten Soviet satellites in Molniya orbits, and these are their crucial communications and early-warning satellites. The situation at present is that the Soviets are more vulnerable to the US ASAT in terms of nuclear war-related missions than is the US to the Soviet ASAT. The most important US systems are way above the range of the Soviet ASAT.

But, by the end of the decade, both sides will have placed more of their most important satellites in geosynchronous orbits. The Soviets, though, for geographical reasons (the effect of the curvature of the earth on line-of sight communications from geosynchronous orbit to the Arctic) will still need to rely on Molniya orbits for some communications functions. Nevertheless, the current US and Soviet ASATs are going to become less capable of inflicting damage on each side's C^3I

facilities. The issue then becomes one of what can each superpower do in the ASAT field to threaten those satellites in semisynchronous and geosynchronous orbits? Here the situation is very assymetrical: the current Soviet ASAT cannot readily be adapted so as to give it a capability against such targets because it is launched into orbit from the ground. To get it to the velocity necessary to place it in geosynchronous orbit would require a rocket booster four times the size of its existing (enormous) booster.[32] More problematically, the Soviet ASAT can only be launched twice a day against any satellite and it is limited in the inclinations it can attack. There are also very few launch sites for it, and currently only about a dozen interceptors.[33] The limitation on the number of launch sites means that only a few could be launched each day: this has led to estimates that it would take a week for the Soviets to destroy those US satellites currently within its reach.[34] The US ASAT, on the other hand, is more flexible, being a direct-ascent weapon. It can be launched against targets of any inclination, at any time. Moreover, there are reports that it is possible to attach a third stage to the launcher so as to give it the capability of attacking satellites in geosynchronous orbits.[35] Nevertheless, there is a generic problem for any attack on satellites in such orbits, which is that any interceptor would take many hours to get to its target, thereby either giving time for manoeuvre or switching to back-up systems. More worryingly, such an attack might be seen as the start of strategic conflict.

If the prospects for any attack on satellites in geosynchronous orbit looks remote with the current generation of ASATs, this benefits the side that relies most on satellites in this orbit, and this is the United States. But both the US and the Soviet Union are working on new forms of ASAT weapons, and these could eventually pose a threat to satellites in geosynchronous orbit. The candidate technologies overlap considerably with those of the US SDI,[36] with the most promising technologies being nuclear-pumped X-ray lasers, directed-energy weapons, kinetic energy weapons, and nuclear explosions. Each of these has advantages and disadvantages, and there is not space available to discuss them in detail. The important point is that such ASAT weapons could pose a serious threat to satellites in all orbits, especially in the case of directed-energy weapons such as lasers and particle beams, since they travel at the speed of light and thus offer no warning. Moreover, because the technologies for such ASAT systems are very closely related to those of SDI, then even a rudimentary SDI system could have significant potential for ASAT missions. The strategic implication is that if such weapons could be developed they would most affect the side that depended on satellites for the management of its nuclear strategy. A clear asymmetry exists here: although more Soviet C^3I satellites are currently threatened by the US ASAT than is the case with the reverse relationship, two factors make the US more vulnerable to future generations of ASAT systems. First, geographical factors make the USSR less reliant on satellite communications than is the case for the US. Not only are their forces less dispersed, but there is also the fact that they have land links to the European theatre whereas NATO has to rely on satellites for its links with the US. Additionally, because Soviet satellites are less sophisticated and reliable, and because they cannot easily provide links to their Arctic forces, the Soviets have tended to use them more as back-up channels of communication.[37] Secondly, the drift in US nuclear strategy we outlined above indicates that the US is moving towards a war-fighting posture that actually increases its reliance on data gained from satellites. That is to say, the US will come more and more to rely on the very systems that advances in ASAT technology would threaten. The paradox can be stated baldly: the Reagan Administration is unwilling to discuss limitations on ASATS, is busy developing technologies for SDI that could have an ASAT role, and yet is adopting a nuclear strategy for which satellites are crucial.

The concurrent development of advanced ASAT capabilities and a nuclear strategy based on notions of war fighting seems likely to create a strategic environment that has very serious destabilising consequences. Such a strategic environment would pose problems for stability at three stages of any crisis: it might make pre-emption more likely, it might make it difficult to control escalation during a conflict, and it would make terminating the conflict more problematic. Of these, the most worrying would be the decline in crisis stability, that is to say, possession of ASATs on both sides might make each side increasingly edgy during a crisis. This is because use of ASAT weapons against early-warning and communications satellites would obviously benefit most the side which attacked first. One does not need to impute an intention to start strategic war to obtain this conclusion; in a crisis, the very fact that each side had weapons that could disable crucially important satellites would increase the incentives to get in first. This becomes more destabilising if we imagine strategic war to be the result of a drift towards conflict elsewhere, for example in Europe. In such conventional conflicts there is a danger that each superpower would be looking for evidence of a forthcoming strategic attack. Not only would the possession of ASATs heighten the perceptions of insecurity, resulting in a feeling that it might be better to get in first, but if one side was to attack satellites in connection with the course of the conventional conflict, this might be seen as evidence of a forthcoming strategic attack. Laser ASATs would be particularly destabilising in this regard since they would heighten fears of a decapitation attack on political and military leaderships. As Gottfried and Lebow point out, during a crisis as the alert status of strategic forces is raised this results in a diffusion of nuclear launch authority as a precaution against decapitation.[38] If the other side observed this, they would need to respond in the same way in order to preserve their control. Inadvertent war might result; ASATs would make this process of delegating control over nuclear forces more likely.

Once the conflict had broken out, satellites would be essential for re-assessing targeting options, and assessing damage. The Soviet Union does not, apparently, accept the US view that protracted nuclear war is possible, believing that if it sees an attack coming the only response is to launch a massive strike.[39] The US is therefore far more reliant on satellites for conducting its nuclear strategy than is the Soviet Union. This must result in a tremendous temptation for the Soviets to disable these satellites early in the conflict. This poses the question of whether this would allow the US to control the escalation process. Finally, the loss of satellites might make it difficult for the superpowers to terminate any conflict; this would also make it very difficult for either side to monitor the other side's compliance with any such agreement.

In short, the possession of advanced ASATs, let alone their use in even a conventional conflict, could make a drift into unintended war more likely. Although there will be attempts to protect satellites from ASATs, the logic of the nuclear age indicates that the advantage will go to the offence, certainly if advanced ASATs exist. A recent lengthy survey by the US Office of Technology Assessment concluded that were the existing unrestrained competition to continue it could lead to severe destabilisation.[40] The problems created by this destabilisation would affect both superpowers, but it is crucial to note that the rule of satellites in US nuclear strategy is far more central than is their role in Soviet strategy. Because the Soviets believe in pre-emption via a massive first attack, and because they do not appear to accept notions of protracted nuclear conflict, they do not ascribe the importance to satellites that is the case in the United States. For the US, it is exactly the link between satellite C^3I facilities and strategic doctrine that makes space-based C^3I systems so crucial. In such a situation, whilst the US might want to develop an ASAT to threaten Soviet space-based C^3I assets (and it is true to say that key Soviet satellites are more vulnerable to the US ASAT than is the case the other way round), it is the US itself that has most to lose in a strategic environment containing advanced ASATs on both sides. The search for military parity, system by system, will result in a situation in which the security of the US is actually decreased, since the Soviets, knowing the central role of satellites in the US offensive force posture, would find these enormously tempting targets in any crisis.

In conclusion, there seems a very powerful case for the US to seek an ASAT arms control agreement. Admittedly the US is not really threatened by the current Soviet ASAT, but if unrestrained deployment continues, the US has more to lose than does the Soviet Union. Fortunately, there does seem to be a short period during which such an ASAT agreement could be reached, but here we have to enter a serious reservation. This is that because the linkage between the technologies for advanced ASAT systems and SDI is so great, there might be very serious verification problems unless SDI is itself subject to agreement. The crucial issue to face is whether the US is better off in a world in which it can threaten Soviet satellites, but have its own satellites threatened, or in a world in which neither side can threaten the satellites of the other side. Under the Reagan Administration the dominant comparison has been between the stages of development and capabilities of the US and the Soviet ASAT weapons; the logic of the argument presented in this chapter is that this is an inappropriate comparison. Because the US relies more on satellites for the implementation of its nuclear strategy than does the Soviet Union, the conclusion of this chapter is that limiting ASAT systems would be in the security interests of the United States. Not only would an ASAT agreement prevent another spiral in the arms race it would also limit the militarisation of space. Back on earth, it would both protect the role of space-based C^3I assets in US nuclear strategy and contribute towards greater stability during conventional and strategic crises. But the Reagan Administration, an Administration that has presided over the further development of war-fighting strategies, looks as if it is pursuing policies that will lead inexorably to the further militarisation of space, and to a decline in crisis stability on the ground.

NOTES

1. Strictly speaking, even this distinction is problematic, since there are considerable overlaps between the candidate technologies for ASAT and SDI systems. A clear example is the close relationship between the US ASAT system currently under test and the ballistic missile defence system known as the Homing Overlay Experiment. Both use similar sensing devices and similar homing devices.
2. *Aviation Week and Space Technology* (28 April 1986) p. 16.
3. Jeffrey Richelson, *The US Intelligence Community* (Ballinger, Cambridge, Mass, 1985) pp. 112–13.
4. *ibid.* pp. 114–15.
5. For some of the problems involved in making these assessments *see* Steve Smith, 'Problems of assessing missile accuracy' *RUSI Journal* **130**(4) (1985) pp. 35–40.
6. *See* Bruce Blair, *Strategic Command and Control* (Brookings, Washington, 1985); Paul Bracken, *The Command and Control of Nuclear Forces* (Yale University Press, New Haven, 1983); Daniel Ford, *The Button* (Simon and Schuster, New York, 1985).

7. The most useful summaries of the history of the US ASAT are contained in: Paul Stares, *Space Weapons and US Strategy* (Croom Helm, London, 1985); Office of Technology Assessment *Anti-Satellite Weapons, Countermeasures, and Arms Control* (Government Printing Office, Washington DC, 1985); John Pike, 'Anti-satellite weapons and arms control', *Arms Control Today* **13**(11) (December 1983) pp. 1, 4–7; Richard Garwin, Kurt Gottfried and Donald Hafner, 'Antisatellite weapons', *Scientific American* **250**(6) (June 1984) pp. 27–37; Union of Concerned Scientists, *The Fallacy of Star Wars* (Vintage, New York, 1984), pp. 179–269; Bhupendra Jasani, 'The arms control dilemma' in his (ed.) *Space Weapons: The Arms Control Dilemma* (Taylor and Francis, London, 1984), pp. 11–21 (*see also* the papers by Marcia Smith (pp. 83–91) and Donald Kerr (pp. 107–25) in that volume); Donald Hafner, 'Anti-satellite weapons: the prospects for arms control' in Bhupendra Jasani (ed.), *Outer Space—A New Dimension of the Arms Race* (Taylor and Francis, London, 1982) pp. 311–23; Paul Stares, 'Reagan and the ASAT issue', *Journal of International Affairs* **39**(1) (1985) pp. 81–94.
8. For an excellent survey of these programmes *see* Paul Stares, *Space Weapons and US Strategy op. cit.* Ch. 6.
9. The Soviet system has had a very mixed record: initially, equipped with a radar homing interceptor, it had a 30 per cent failure rate; the later version, using a heat-seeking sensor system, experienced a 100 per cent failure rate. It also has considerable operational limitations, having to wait until the target satellite approaches the launch site, which happens only twice a day, and being limited to inclinations of between 62° and 65°: it cannot therefore attack US satellites on lower inclinations.
10. For details of this *see Aviation Week and Space Technology* (7 October 1985) pp. 18–19.
11. *See* Paul Stares, *Space Weapons and US Strategy op. cit.* Ch. 9.
12. *See* Note 9 above.
13. *Arms Control Today* **15**(17) (September 1985) p. 15.
14. *See Science* (24 January 1986) pp. 331–2; *Aviation Week and Space Technology* (23 December 1985) p. 19; *Arms Control Today* **15**(9) (November/December 1985) pp. 20–1.
15. *Science* (24 January 1986) p. 33.
16. *Arms Control Today* **15**(9) (November/December 1985) p. 20.
17. *See* Jeffrey Richelson, 'PD-59, NSDD-13 and the Reagan Strategic Modernization Program' *Journal of Strategic Studies* **8**(2) (1983) pp. 125–46; Phil Williams, 'Deterrence, warfighting and American nuclear strategy', *ADIU Report* **5**(1) (1983) pp. 1–5; Lawrence Freedman, *The Evolution of Nuclear Strategy* (Macmillan, London, 1981); Gregg Herken, *Counsels of War* (Knopf, New York, 1984); Desmond Ball, 'Counterforce targeting: How new? How viable?' *Arms Control Today* **11**(2) (1981) pp. 1–2, 6–9; Desmond Ball, *Can Nuclear War be Controlled?* (IISS, Adelphi Paper 169, London, 1981); Colin Gray, 'Nuclear strategy: The case for a theory of victory', *International Security* **4**(1) (1979) pp. 54–87; Colin Gray, *Nuclear Strategy and*

Strategic Planning (Foreign Policy Research Institute, Philadelphia, 1984).

18. Quoted in Richelson, 'PD-59, NSDD-13 and the Reagan Strategic Modernization Program' *op. cit.* p. 127.
19. Blair *op. cit.* p. 26.
20. Ball, 'Counterforce targeting: How new? How flexible? *op. cit.* pp. 6–7.
21. *ibid.* p. 7.
22. For a clear discussion of the ideas behind this *see* Walter Slocombe, 'The countervailing strategy' *International Security* **5**(4) (1981) pp . 18–27.
23. Cited in Blair *op. cit.* p. 27.
24. *ibid.* pp. 27–8.
25. For a strong statement supporting the adoption of a war-fighting strategy *see* Colin Gray and Keith Payne, 'Victory is possible', *Foreign Policy* No. 39 (1980) pp. 14–27.
26. *Department of Defense Annual Report Fiscal Year 1984* (Government Printing Office, Washington, 1983), pp. 51–3.
27. *See* Blair *op. cit.* p. 246.
28. For discussions of these implications, *see* Union of Concerned Scientists *op. cit.* pp. 227–48. Kurt Gottfried and Richard Ned Lebow, 'Anti-satellite weapons: weighing the risks', *Daedalus* **114**(2) (1985) pp. 147–70; Office of Technology Assessment *op. cit.* Thomas Karas, 'Military satellites and war-fighting doctrines' in Bhupendra Jasani (ed.), *Space Weapons—the Arms Control Dilemma op. cit.* pp. 43–55.
29. These functions are discussed for the US and the USSR in, respectively, Jeffrey Richelson, *The US Intelligence Community op. cit.* Chs. 7 and 8 and his *Sword and Shield* (Ballinger, Cambridge, Mass, 1986) Ch. 5.
30. The inclination of the satellite is also important, and represents a serious limitation on the Soviet ASAT.
31. This information is based on data presented in Bhupendra Jasani, *Space Weapons—The Arms Control Dilemma op. cit.* Table 1, p. 6.
32. Union of Concerned Scientists *op. cit.* p. 199.
33. John Pike, 'Anti-satellite weapons and arms control' *op. cit.* p. 7.
34. Kurt Gottfried and Richard Ned Lebow *op. cit.* p. 151.
35. Union of Concerned Scientists *op. cit.* p. 203.
36. These are discussed in Office of Technology Assessment *op. cit.* pp. 62–75.
37. Union of Concerned Scientists *op. cit.* pp. 194–5.
38. Kurt Gottfried and Richard Ned Lebow *op. cit.* pp. 162–3.
39. *See* Stephen Meyer, 'Soviet military programs and the "new high ground" ', *Survival* **25**(5) (1983) pp. 204–15.
40. Office of Technology Assessment *op. cit.* p. 130.

7 The Militarisation of Space Through Soviet Eyes

Stephen Shenfield

INTRODUCTION

Soviet speculation concerning the militarisation of space[1] is as old as the space age itself. That the scope of military operations may in the future extend to outer space was noted in 1962 by Marshal Sokolovskii in his *Military Strategy*.[2] The development of ASAT weaponry in recent years has brought this future nearer, and since 1983 Soviet commentary has focused on the American Strategic Defence Initiative as the catalyst of a possibly imminent space militarisation. This chapter is devoted to a review and assessment of such commentary.

To place the issue of space militarisation in its context, it should be borne in mind that there is no one-to-one correspondence between classifications of weapons systems by site of deployment (land, sea, air or space), by site of location of targets, by envisaged role (ASAT, ABM etc) and by technological basis (nuclear, laser etc). There is a tendency in Western debate about SDI to conflate the site category of space weapons, the role category of ABM and the technological category of directed-energy weapons. Thus, Soviet directed-energy research is often interpreted in terms of a 'Soviet SDI'. However, Soviet strategists anticipate the use of directed-energy weapons not only in space but also in Earth battlefields.[3] As we shall see, they may not view the ABM role as primary for space weapons. And it is likely that they regard land, not space, as the optimal site for ABM deployments.

Therefore, our study of Soviet views of the militarisation of space does not cover either the question of Soviet antimissile defence[4] or that of Soviet directed-energy weapon programmes, taken as a whole.

Any analysis of Soviet strategic views which relies on open sources must answer, however briefly, the objection that such sources reveal 'not what the Soviets think, but what they would like us to think they think'. Some analysts, like Richard Pipes,[5] discount *all* open sources on these grounds, advocating that real Soviet views be assessed on the basis of closed military sources and actual deployments. Others, such as Stephen Meyer,[6] distinguish between those open sources which do serve a propaganda function and those, especially military sources, which do not, discounting only the former.

In the field of space warfare, there are as yet neither Soviet deployments nor an elaborated Soviet strategic doctrine, so that we have no alternative to extracting whatever insight we can from Soviet sources. Even propaganda designed to mislead may reveal certain deep assumptions and concerns of the propagandist. Moreover, in this field 'propaganda' and 'internal' sources are consistent with one another; there seems no reason to accept one while rejecting the other. Soviet sources may be some kind of mirror, though undoubtedly a selective and distorting one, of real Soviet attitudes. They require critical assessment.

We shall, for example, refer to the report on SDI that has been produced in a variety of editions by a working group of the 'Committee of Soviet Scientists for Peace, Against Nuclear Threat'.[7] The report as published is clearly aimed at influencing Western opinion. On the other hand, according to Vice-President of the Academy of Sciences Academician Velikhov, the report includes 'only a small proportion of the materials and calculations prepared by various members of the working group and its subgroups'.[8] Gorbachev has stated, apparently in reference to the same study, that 'the Soviet leadership gave appropriate instructions to competent organisations and scientists at one time' to consider possible Soviet responses to SDI.[9] It seems likely that the propaganda exercise is a selective by-product of an important Soviet policy study, and it is in this light that we judge the contents of the report.

Besides sources published in the USSR, we make use of an assessment of the strategic consequences of space militarisation by the pseudonymous Soviet expert on international and military affairs Viktor Olenev, published in the British Soviet Studies journal *Detente*. Olenev's analysis, while not inconsistent with more official Soviet sources, explores aspects of Soviet thinking on space militarisation merely hinted at or ignored in the open sources.[10]

Our analysis begins with an examination of Soviet assessments of the technical feasibility of space weapons systems, both for the USA and for the USSR. Soviet commentators envisage three possible roles for space weapons: they 'can be used against missiles, against satellites and against targets on Earth'.[11] We accordingly consider each of these roles in turn from the Soviet point of view. Then we discuss Soviet policy on space militarisation and its perspectives. In conclusion, we sum up what seem to us the most important points.

SOVIET ASSESSMENTS OF TECHNICAL FEASIBILITY

The views expressed on the technical feasibility of SDI by the Soviet scientists who have been studying the question show curious shifts and inconsistencies. When Velikhov spoke on SDI at the Pugwash conference in August 1983, he dwelt upon the exacting requirements even of a chemical-laser system, 'the best candidate for a space-borne system', but added that 'one must never underestimate people's ingenuity, and one may envisage some improvement (in cost effectiveness) if one were prepared to spend a billion dollars'.

Yet after discussing the strategic implications of SDI, he dismissed the relevance of his own speculations: 'I do not believe that these considerations are very practical, because the technical question is clearly more important. Maybe the most important practical consequence is simply that even an illusion of the possibility of realising such a system can divert our attention from disarmament.'[12]

The 1984 edition of the scientists' report states that SDI has not yet progressed from laboratory prototypes of individual elements to technical development of the system as a whole. There follows the forecast: 'On the basis of comparison with Western programmes of similar scale, we may conclude that from the beginning of development work to putting the system into operation will take at least ten years.' The forecast that weapons may be deployed in space within 10–20 years has recurred in Soviet commentaries, including a speech delivered by Gorbachev in November 1985.[13]

Later editions of the scientists' report, however, retreat from any projections of time-scales, omitting the sentence cited above. The ten-year estimate is also implicitly rejected by one strategic expert associated with the report, who writes that 'even a trial of a laser platform would require significant efforts and expenditures over a number of years', while problems would be much greater for a whole system: 'It becomes understandable why many authoritative scientists *in the USSR*, USA and other countries express serious doubts concerning the feasibility of a space ABM system in principle.'[14]

The clearest expression of Velikhov's scepticism comes in his introduction to a book on the nuclear winter published in 1985. Stressing the unwieldy size of existing directed-energy installations, he argues that by 'a realistic assessment of any laser technology devisable in the foreseeable future . . . lasers do not hold much promise', and particle-beam weapons are even less promising: 'To be sure, the people who invent weapons systems have an unlimited capacity for fantasy. But sensible people should remember that projects of this kind depend . . . on new technical solutions and design of instrumentation that are well beyond the imagination, let alone the capability, of scientists today.'[15]

There is then no consistent Soviet assessment concerning technical feasibility. It appears that the natural scientists are more conscious of the practical difficulty of weapons deployment in space (Velikhov, for instance, formerly directed the Institute of Atomic Energy laboratories at Troitsk, where military lasers are developed). They therefore tend to take a more sceptical view of future possibilities than that typical of strategic analysts who are not natural scientists, inclined towards the worst-case assumption that technical problems can to a large extent be overcome. The 'fantasts' at whom Velikhov rails do not necessarily all live in the USA.

Thus, a pamphlet issued by the Military Publishing House states, without a shadow of equivocation, that 'modern science and technology make it possible to develop' laser, particle-beam and other new types of weapons deployable in space.[16] Olenev (who has a military background) warns of 'fully battle-ready space-force complexes' by the year 2000, on the basis of an analogy with military aviation, which took only 25 years to develop from a few intelligence planes to become a key factor in World War Two.[17]

If reliance on historical experience generates expectations of new American technological successes, it also produces a certain degree of confidence in Soviet ability to meet the American challenge, for 'historical experience shows that advantage is temporary and indecisive'.[18] The USSR was also, of course, the first to put a satellite into orbit and a man into space. As factors facilitating the military mastery of space by the USSR, should it prove necessary, Olenev cites the high educational level of the Soviet people, the raw-material self sufficiency of the Soviet economy and its adaptation to extreme concentration on set goals, and the dynamism infused into the system by 'the new generation of young, energetic and qualified leaders'.[18] In this connection, we note that many in the Soviet elite rate Soviet technological potential more highly than some Western observers do, because they attribute the difficulties of Soviet technical advance not to the intrinsic nature of the Soviet system but to less basic factors like the passive and incompetent leadership of the Brezhnev era.

At the same time, a facade of self-assurance does serve a deterrent function, and some signs of anxiety are visible behind it. 'The elements of unpredictability, chance and spontaneity in scientific discovery', write two Soviet analysts of the arms race, 'induce in the opposing sides uncertainty with regard to their strategic plans and the real possibilities of qualitatively raising their military potential'.[19] Even a temporary disadvantage which gives the USA 'a partial illusion of invulnerability' is fraught with the danger that the USA may be tempted 'to catch the moment for a first strike with "acceptable" losses'.[20]

Finally, Soviet expectations of technological advance no doubt vary greatly from one area of technology to another. Soviet planners may expect to remain in the forefront of laser development, for instance, but can hardly hope rapidly to overcome their lag in computers, and sensor and control systems.[21]

SPACE WEAPONS AS AN ABM DEFENCE

The Soviet critique of SDI as an ABM system is very similar (though not identical) to that of SDI's Western critics.[22] Inevitably vulnerable to a host of counter-measures, SDI cannot conceivably serve its ostensible purpose, but it will generate pre-emptive pressures on both sides and heighten the risk of accidental nuclear war.

Soviet commentators envisage both passive counter-measures to SDI, designed to overwhelm or evade the system or protect missiles against it, and active ones, which would threaten to destroy or put out of action space-based installations. The scientists' report notes that the technologies required for countermeasures are in a much more advanced state than those required for SDI itself, with many of them ready for us. 'An effective network of means of counteraction' can be rapidly set up at a cost of only 1–2 per cent of that of the system being counteracted.[23]

Passive counter-measures mentioned include: further build-up in the number of strategic missiles, and in the number of warheads on existing large ICBMs; diversification of the nuclear arsenal, with increased reliance on Cruise and sea-based missiles and on lighter land-based missiles with fast-burn boosters; use of decoys; encasing of missiles with laser-absorbing or reflecting material; masking missile launches by smokescreens; and ordering missile launches over time in such a way as to force the lasers constantly to redirect themselves in haphazard fashion.

Active counter-measures contemplated are: relatively small ballistic missiles based on land or sea, in air or space; satellites placed in orbits close to those of the laser platforms and equipped with short-range means of destruction ('space mines'); land-based lasers; and obstacles (e.g. 'clouds' of heavy spheres) laid along the trajectories of laser platforms to collide with them. Thus, the USSR intends to counter SDI with its own 'space strike weapons'.[24] SDI, as a parallel system under threat of attack from a series system, would be extremely vulnerable to partial or complete destruction.[25]

It is in this context that the Soviet perception of SDI as a 'Strategic Offence Initiative'[26] must be understood. It is inconceivable that SDI could be robust and effective enough to provide worthwhile protection against a Soviet first strike, which would be accompanied by an attack on the SDI platforms. However, SDI 'may create illusions concerning the feasibility of defence against a second, retaliatory counterstrike'. Such illusions would be based not only on the expectation that a retaliatory strike would be less powerful and co-ordinated than a first strike, but also on the calculation that infliction of a first strike on the USSR would impede the taking of effective counter-measures against SDI.

In a crisis situation, fear of losing SDI to a Soviet attack might tempt American leaders to launch a pre-emptive strike. The perception of Soviet leaders that their American counterparts were under this temptation would tempt them in turn to pre-empt. The vulnerability of SDI 'makes it even more destabilising'.[27] (A point neglected in Soviet commentary is that the vulnerability of *Soviet* space weapons would induce a 'use or lose' dilemma with analogous effects.)

Another reason why SDI space platforms have much greater potential in conjunction with a first strike than in defence against a first strike is that they would be within range of Soviet missiles over only 5 per cent of each orbit before going over the horizon.[28] SDI is viewed, moreover, as one element in a broader drive by the USA to enhance its first-strike capability, other elements being the expansion of American strategic forces and the deployment of Pershing II missiles in Europe. The refusal of the USA to follow the lead of the USSR in undertaking not to use nuclear weapons first is also cited in this connection.[29] Deputy Defence Minister Marshal Petrov asks why, if the USA seeks a transition to a defence-dominated world, the Pentagon is intensively working on means to counteract a possible future Soviet SDI.[30]

There has been a debate in the American literature about the possibility of negotiating measures which would assure the USSR that SDI does not form part of a first-strike strategy. David Rivkin, for example, argues that this might prove feasible, while Stephen Meyer denies that Soviet perceptions are ever likely to evolve in this direction.[31] The decisive consideration here is that, even if the USSR were to be assured that the USA does not seek a first-strike capability by means other than SDI, it would still see the strategic logic *inherent* in space weapons as an offensive one.

Another destabilising characteristic of space weapons, in the view of Soviet writers, is their vulnerability to accidental collisions which might be automatically interpreted as deliberate attacks. Gorbachev has speculated that a separated fragment of a test missile might collide with a space weapon subsystem. 'All computers would be switched on, while politicians would not be able to do anything sensible.'[32] '. . . Decisions, irreversible in their consequences, would be taken by computers, without participation of human mind and political will, . . . Such a development could result in a universal catastrophe—even if the initial impulse were an error, miscalculation, or technical malfunctioning of sophisticated computer systems.'[33]

The *leitmotiv* of Soviet comment on the militarisation of space is that it will generate unpredictability, indeterminacy, 'strategic chaos' and, in general, loss of control. Military space developments 'may in the future find application in the most unexpected directions'.[34] 'The whole system for assessing the strategic balance' will be greatly complicated and confused, making it difficult to compare the increasingly asymmetric strategic forces of the two sides.[35] Neither side will be sure of its own capabilities or of those of its adversary, and 'even if the two sides have roughly equivalent nation-wide ABM systems, . . . small changes in their efficiency would inevitably and substantially upset military–strategic parity and destabilise the strategic situation'.[36]

Strategic indeterminacy would in itself exacerbate mutual distrust and suspicion. In short, as Gorbachev reportedly asked Reagan at the 1985 Summit: 'Who can control it? Who can keep a grip on it?'[37]

Soviet propagandists may be suspected of exaggerating the real degree of concern for strategic stability on the part of the Soviet leadership, especially when one recalls such destabilising Soviet actions as the renewal in the late 1970s of the then dormant race in antisatellite weapons.[38] On the other hand, there is also ample evidence that the Soviet leadership do fear 'the uncertainty which attends loss of control over both events and forces as the major destabilising factor in a crisis'[39]—for example, their proved reluctance to raise nuclear alert levels. It is not unreasonable to credit their warnings about the dangers of space militarisation with some measure of sincerity.

SPACE WEAPONS AS ANTISATELLITE WEAPONS

Whatever the apparent inconsistencies of Soviet policy on the question of antisatellite weapons, all Soviet commentary recognises the stabilising function of surveillance satellites and the threat to stability posed by the ability of ASAT to 'blind' strategic forces.[40] Moreover, Soviet analysts, like Western critics of SDI,[41] consider ASAT technically the most feasible role for space weapons, and one which would have a destabilising effect on the strategic situation in the quite near future, irrespective of the ultimate results of SDI.[42] Those scientists and other civilian analysts who take a sceptical view of the feasibility of more demanding roles see the threat to satellites, and to the space weapon objects of the other side, as the *main* danger even in the longer term: 'The most dangerous directions of the militarisation of space are those connected to realising the possibility of direct military operations in space . . .—from interference with the normal work of (the other side's) space apparatuses up to their destruction.'[43]

Velikhov asks whether, in view of the technical infeasibility of SDI, it is 'worthwhile to be worried about the prospects of SDI deployment'. He gives three reasons for a positive answer. Firstly, an effective ASAT capability will emerge even at the early R&D stages of SDI. Secondly, there is an '*X* factor': 'although not feasible in its full scope', SDI marks the beginning of a dangerous 'spiral' of an arms race in 'varied and many space weapons systems' with 'totally unpredictable consequences'. It 'opens a Pandora's Box'. Thirdly, SDI is pursued as an alternative to disarmament and undermines the 1972 ABM Treaty, the basis of arms control.[44] In this analysis, ASAT is the *only* specific threat identified.

Although in 1983–4 the USSR combined its opposition to SDI with a campaign for a moratorium on the testing and deployment of antisatellite weapons, this campaign was suddenly abandoned at the beginning of 1985. The most plausible explanation for the change is that the USSR wishes to reserve the right to develop its ASAT capability as an active counter-measure against American SDI space platforms.[45] However, as ASAT capability is an inevitable outcome of the early stages of the SDI programme, the USSR thereby gives up the diplomatic tool which might have given it the best prospects of stopping SDI. This casts at least some shadow of doubt on the strength of the USSR's commitment to prevent the militarisation of space.

SPACE WEAPONS AS WEAPONS AGAINST TARGETS ON EARTH

Soviet analysts, with the exception of those few who rule out the technical feasibility of any but the ASAT role, emphasise that 'space strike weapons' can be directed not only against missiles in flight and space objects, but also against targets on Earth:

> In time, orbital battle-stations may be turned directly into threatening offensive arms. They will become capable of destroying from space the military–political leadership, armed forces, population, industrial objects, infrastructure, varied targets on land and sea, in air and space. And this terrible weaponry will permanently hang literally over everyone's head and possess . . . an instantaneous speed of action.[46]

Commonly mentioned targets are command, control and communication centres, missiles at launch sites, aircraft on airfields, power stations, oil refineries and ports.[47] Space lasers could set off huge fires to destroy cities and large crop areas.[48]

A distinction is observable between military and non-military sources in the relative stress put on the space-to-earth weapon (STEW) role of space weapons. While almost all sources point out that there is 'no guarantee' that space weapons may not be used to destroy Earth targets,[49] civilian sources (such as the scientists' report) tend to mention the matter *after* discussing the ABM and ASAT roles, while many military sources dwell *first* on the STEW role. For example, in the chapter of the pamphlet on 'Star Wars' issued by the Military Publishing House which substantiates the offensive nature of SDI,[50] the STEW capability of SDI is the first argument adduced.

Given the prominence of this theme in the Soviet literature, it is astonishing that it has received so little attention in Western analyses of Soviet thinking. Note should be taken of the fact that the STEW threat dominates the language and imagery of Soviet discourse about space militarisation. 'Space strike weapons' are 'a space sword raised over the world'.[51] They 'hang or fly' over all countries, 'able to reach and strike at any target around the world'.[52] Volkogonov draws a parallel between the Martian threat from space in H. G. Wells' *War of the Worlds* and the threat from space which may be posed by Earthlings.[53] 'Space must not become . . . a source of death and destruction', states the new Party Programme.[54] And Gorbachev has pinpointed the contrast between Western discussion of space weapons in the context of strategic defence and the Soviet perception: 'The American side is reluctant to admit that SDI means bringing weapons to space. Weapons. They would fly over people's heads in waves—American and Soviet weapons. We would all watch this sky and expect something to fall from there.'[55]

A variety of arguments are given in the Soviet literature for regarding space weapons as having a directly offensive function. First of all, the STEW role is believed to be technically less difficult than the ABM role: 'It is no secret to a specialist . . . that laser weapons can be used more effectively and simply from space against ground targets than against flying missiles'.[56] Secondly, reference is often made to the currency in US Air Force circles of the idea that the power which first achieves supremacy in space will thereby dominate the Earth as well.[57] Lastly, signs of a hidden STEW orientation are sought in the American SDI programme. Thus, the fact that 'American publications on SDI have lately emphasised kinetic-energy weapons' is attributed to the special potential of such weapons for use against ground objects.[58]

We might also view the emphasis on the STEW role as a natural product of the traditionally offensive cast of Soviet military thought. The idea that offensive weapons could be placed in space is far from new in the USSR. Soviet leaders in the early 1960s envisaged the use of satellites as missile platforms. In 1961 Khrushchev boasted that 'we placed Gagarin and Titov in space, and we can replace them with other loads that can be directed to any place on Earth'. In 1963 Chief of Strategic Missile Forces Marshal Biryuzov claimed that 'it has now become possible at a command from Earth to launch missiles from satellites at any desired time and at any point in the satellite trajectory'.[59] In fact, of course, this did not prove an effective launching mode. Nevertheless, Soviet analysts have now revived the idea of orbiting nuclear weapons as a conceivable countermeasure to SDI: the high re-entry speed of missiles launched from orbit would help them penetrate the terminal phase of the adversary's BMD, though it might not be possible to make them very accurate.[60]

Soviet analysts appear to view the STEW role of space weapons in two different strategic contexts. First, 'simultaneous strikes on Earth targets', including unlaunched missiles, could form part of a first nuclear strike.[61] The 'instantaneous' nature of the space component of the attack would be particularly advantageous for surprise pre-emption and decapitation. Weapons in near-Earth orbit could come within 100–500 km from their targets, so that flight times would be minimal (in the case of nuclear weapons) or zero (in the case of directed-energy weapons).[62]

Second, space strike weapons 'can appear over the territory of any State at practically any moment and create a tangible threat to its security': such threats could be used to exert pressure on individual States, especially non-nuclear ones, outside the context of global war.[63] Anatoly Chapis of the Institute of World Economy and International Relations explains:

> This means that the USA will have obtained a powerful means of pressure not only on the Soviet Union. 'Star' weapons would flash over practically any

> point of the globe at regular intervals. While at present intervention in a regional conflict requires the transport of considerable forces and equipment to the area concerned, the deployment of weapons in space removes this difficulty. A space fist can be used at any moment and against practically any country.[64]

While the open literature notes only the impact of this new capability on American power, Olenev in his unofficial analysis also speculates about a corresponding enhancement of Soviet power. Olenev is perhaps one of those whom Vasilyev and Oznobishchev have in mind when they write that 'some specialists regard (the space strike weapon) as a "superweapon", capable of radically changing the correlation of forces on the world arena'.[65] Olenev argues that, as only the USA and the USSR can compete in space, the militarisation of space would consolidate the dominance of the two superpowers over all other states, reversing the recent trend toward multipolarity in the international system. While the superpowers would both gain in power at the expense of the rest of the world, the USSR would gain the most, finally attaining full parity with the USA as its remaining disadvantages on Earth (inferior naval forces to those of the West, smaller political and economic influence, etc.) were neutralised by its power in space: 'Space forces . . . will give the USSR a permanent global military–political presence. They will allow it to be present at all times and places, on an equal basis with the USA, where we are now practically absent.'

Olenev puts his stress on what is detectable in open sources as a minor theme: the USA may be seeking not absolute supremacy in space, but rather to negotiate a controlled arms race there under a superpower pact directed against third parties. Gorbachev said at Geneva: 'There has been a kind of invitation from the American side: let us see, let us make it out, let us talk not of how to prevent space being militarised but of what kind of weapons to take into space.'[66]

Olenev also interprets leadership statements about the need to take account of the legitimate interests of non-superpower States as criticisms of the idea of a superpower pact.[67] In his view, Soviet policy is subject to the pressure of conflicting interests. As a 'national State' concerned with averting the risk of war and raising the living standards of its people, the USSR is led to oppose the militarisation of space, while as a 'great world power' concerned with its relative military might in the world, the USSR is tempted to accept the 'invitation' of the USA to join it up in space.

SOVIET POLICY ON SPACE MILITARISATION AND ITS PERSPECTIVES

The policy choices facing the USSR are of two kinds: what priority to give to efforts to prevent, or if necessary reverse, the militarisation of space, and how to develop its own military space programme to meet the contingency of space militarisation not being prevented. These two tasks, though to some extent compatible, do come into conflict with one another, as we have shown in the case of ASAT policy, and the tension between them can be expected to increase over time.

Soviet policy orientations appear to vary along two dimensions: feasibility and desirability of conceivable developments. If, like Velikhov, one sees only the ASAT role as feasible, then one can be in little doubt that the balance of advantage for the USSR lies against space militarisation. If ABM and/or STEW roles are regarded as feasible, then—as Olenev argues—the balance is less determinate, and one's attitude will depend on various speculative calculations of risks and opportunities as well as on value judgements. Implicit in Olenev's analysis is the probable existence of 'Young Turks' in the Soviet military, inclined to accept the risks and costs of space militarisation for the sake of the enhancement of the internal and global status of the armed forces that they expect from it.

We should allude briefly to a range of non-military considerations which bear on Soviet attitudes to space militarisation. The likely economic burden is a powerful disincentive to accepting an arms race in space, which would 'very seriously slow down the rise in living standards of the Soviet people'. On the other hand, Soviet commentators deny that this burden would be insupportable. That this need not always be a propaganda stance is suggested by Olenev's argument that the space race will 'to a certain extent have a positive effect on our economic and scientific–technical advance', speeding up the computerisation of the economy.[68] Mikhail Agurskii, who himself worked in a Soviet military research institute, draws attention to coded debate in the Soviet literature between the military 'technocrats' who hold that the arms race stimulates technical progress and the traditional bureaucrats who repudiate this idea.[69]

Olenev also discusses the impact of internal Soviet politics. 'The threat from space' serves the interests of 'neo-conservatives', ready to exploit it for the purpose of reinforcing popular discipline and blocking social reform. At the other end of the spectrum, it is the 'progressive' forces, committed to democratisation and detente, who have the strongest interest in preventing the militarisation of space. A space race 'will strengthen Communist ideology in the USSR' and arouse 'hatred of the West' and 'war hysteria'.[70]

There have been apparent signs of disagreement within the Soviet elite concerning whether to restrict a Soviet response to SDI to 'counter-measures' or to pursue more ambitious goals. For example, Anatolii Gromyko and Vladimir Lomeiko write that Soviet counter-measures 'might not necessarily fully correspond to the actions of the US, because the Soviet Union . . . is far from the thought of extending the arms race to space'.[71] By contrast, Chief of General Staff Marshal Akhromeyev accepts no constraints on the Soviet response: 'Nothing will be left to us but to take answering measures both in the field of offensive arms and in that of other arms, not excluding defensive arms, among others space-based.'[72]

Gorbachev's public position implicitly endorses the counter-measures-only strategy outlined in the scientists' report: 'We can say that our response would be effective, less expensive and could come within shorter time-limits'.[73]

However, the apparent contrast between the options of a full symmetric response to SDI and a minimal asymmetric response becomes less striking when it is recalled (Section 3) that 'countermeasures' may well include the deployment of Soviet space weapons to threaten American ones. Gorbachev has stated that the USA 'cannot hope to have a monopoly' of weapons in space.[74]

Some Western analysts fear that the USSR might

contemplate taking active measures to disable or destroy American SDI platforms at any time. It is suspected that a claim to the right to do this is implicit in the Soviet argument that the deployment of weapons by one State in outer space above the territory of other States constitutes a breach of their national sovereignty.[75]

If a relatively ambitious military space programme is eventually adopted, it is unlikely to imitate the American programme. The scientists' report predicts that SDI will further accentuate the existing asymmetry in the force structures of the two sides, for the development of space systems, 'like that of strategic offensive forces, will probably follow different routes' in the USSR and the USA.[76] Defence Minister Marshal Sokolov has said that: 'the Soviet Union will choose the modes of action that best meet the interests of its defence capability, and not those which the people in Washington might urge it to take'.[77]

Soviet military men, writes Olenev in references to Reagan's offer to share SDI technology with the USSR, do not wish to receive ready SDI designs from the USA, as they cannot agree 'to trail along behind the US Space Forces'.[78]

The open sources give us little if any guidance on the question of how the Soviet 'route' in space is likely to differ from the American. However, we would suggest the hypothesis, on the basis of the way Soviet commentators describe the militarisation of space as a general prospect, that the USSR may give higher priority to the targeting of Earth objects from space and a relatively low priority to space-based ABM defence (while concentrating, in the event of collapse of the ABM Treaty, on land-based ABM defence). Further investigation is needed on these matters.

But no doubt the most important aspect of Soviet policy in the near future is the intensity and duration of the efforts of the USSR to prevent space militarisation. For some years this goal may represent the main thrust of Soviet policy in this field. However, to the extent that such efforts are seen to fail, there may well follow an adjustment to the new situation as a 'fact of life'. The emphasis would then shift to the goal of deriving from the arms race in space the greatest possible gain in strategic power for the USSR. The internal forces still opposed to the militarisation of space and aspiring to reverse it would tend to become weaker, while those forces which welcome the challenge of space power would grow stronger.

This prognosis is consistent with the three-stage pattern of response to SDI predicted by Olenev.[79] The first stage, to about 1990, is 'a period of waiting, consolidation, reaction and preparation', in which there is still substantial hope that space militarisation will be halted: social–democratic governments may come to power in Western Europe, and Reagan's successor may be much less committed to SDI. Passive counter-measures to SDI are undertaken; the missile force is diversified to evade SDI (development of lighter missiles, 'a gradual transfer of the centre of gravity of nuclear forces from land to sea'). There is preparatory work on space weapons, but they are not deployed.

During the second stage 'combat Space Force objects', perhaps relying on conventional means of attack, are deployed, as well as elements of more advanced systems. Counter-measures now include 'systems for disorienting anti-missile defences'. 'A certain restraint will be shown in the expectation of an end to the militarisation of space'.

'Restraint' no longer characterises the third stage, around the turn of the century, when the full 'complex of Space Forces' emerges together with 'a new military strategy and a new structure of armed forces and armaments in space, in the air, on the sea and on land'.

SUMMING UP

We have shown that space weapons are envisaged in Soviet commentary in three roles. The role nearest in time, and the only role specifically expected by those analysts most sceptical about the technical feasibility of complex space systems, is the targeting of space objects—the satellites and space weapon elements of the adversary. The ABM role is viewed, on the grounds of the strategic logic inherent in space weapons, as potentially effective only in the context of a first nuclear (and space) strike. A prominent position in Soviet thinking, especially in that of military men, is occupied by a third role for space weapons—the targeting of objects on Earth, including C^3 facilities and unlaunched missiles.

Soviet policy on space militarisation is subject to the pressure of conflicting interests. It is seen as fraught with the risk of strategic chaos and instability and loss of political control, increasing the probabilities of pre-emptive strikes (by either side) and of accidental nuclear war. It would also be an economic burden, though a supportable one. On the other hand, space may enhance the strategic power of the USSR, ensuring it effective parity with the USA and a position of dominance over non-superpower States.

Soviet efforts to prevent (and later to reverse) the militarisation of space can be expected to continue, but to the extent that they are seen to fail they may well lose intensity with the passage of time, as the USSR adapts itself to the task of gaining maximum advantage from the new situation.

NOTES

1. In Soviet usage, it is the deployment of weapons, and not just satellites, in space that constitutes its militarisation.
2. V. Sokolovskii, *Voennaya strategiya* (Moscow, 1962) p. 233.
3. Marshal Ogarkov foresees the emergence, beyond the next decade, of battlefield weapons 'based on new principles of physics'—i.e. directed-energy weapons (*Istoriya uchit bditel'nosti*, Moscow, 1985).
4. For a short assessment of the Soviet ABM programme, *see* J. Pike in E. P. Thompson (ed.), *Star Wars* (Penguin, Harmondsworth, 1985) p. 50. Alternative interpretations of Soviet thinking on BMD are given by S. Stevens and R. L. Garthoff in A. B. Carter and D. N. Schwartz (eds.) *Ballistic Missile Defense* (The Brookings Institution, Washington DC, 1984), pp. 182–220, 275–329.
5. See 'A Debate Between Raymond Garthoff and Richard Pipes on Soviet Nuclear Strategy', p. 169 in P. E. Haley, D. M. Keithly and J. Merritt (eds.), *Nuclear Strategy, Arms Control and the Future* (Westview Press, Boulder, Col., 1985).
6. S. M. Meyer, 'Soviet strategic programmes and the US SDI', *Survival* (November–December 1985) p. 274.
7. We refer to two of the Soviet Russian-language editions: (a) the rotaprint *Strategicheskie i mezhdunarodno-politicheskie posledstviya sozdaniya kosmicheskoi protivoraketnoi sistemy s ispol'zovaniem oruzhiya napravlennoi peredachi energii* (IKI AN SSSR: Moscow, 1984); and (b) the document of the same title published in *SShA* 1985, **11**, pp. 112–27.

There is an earlier rotaprint edition (1983) and a more recent edition issued as a brochure. We refer also to the most recent of a number of Soviet English-language editions: (c) Ye. Velikhov, R. Sagdeev and A. Kokoshin (eds.), *Weaponry in Space—the Dilemma of Security* (Mir Publishers, Moscow, 1986). For a shortened translation of the report, *see Survival*, March–April 1985, pp. 83–9, and for a summary of it—*Detente* **3** (May 1985) pp. 10–12.
8. Introduction to edition (b) of Note 7, p. 114.
9. 'Mikhail Gorbachyov's press conference in Geneva', *Soviet News* (27 November 1985) p. 441.
10. V. Olenev, 'The threat from above: the USSR and the Militarisation of Space', *Detente* **5** (Winter 1986) pp. 2–6. This is a shortened translation of Olenev's draft, the full text of which is available from this writer. *See also* M. Frankland, ' "Sky's the limit" for Soviet gains in global power', *The Observer* (23 March 1986) p. 17.
11. As Note 9.
12. E. P. Velikhov, 'Outer space warfare' in J. Rotblat and A. Pascolini, *The Arms Race at a Time of Decision—Annals of Pugwash 1983* (Macmillan, London, 1984) pp. 19–24.
13. Speech by Gorbachev to Nobel Prizewinners on 13 November 1985 (*Krasnaya zvezda* 14 November 1985). *See also*, for example, V. Nekrasov, 'Moment of truth', *Twentieth Century and Peace* (1986) No. 3, pp. 2–6.
14. A. G. Arbatov, 'Ogranichenie protivoraketnykh sistem—problemy, uroki, perspektivy', *SShA* **12** (1984) p. 16.
15. *The Night After . . .—Climatic and Biological Consequences of a Nuclear War* (Mir Publishers, Moscow, 1985) pp. 27–8.
16. *'Star Wars': Delusions and Dangers* (Voenizdat, Moscow, 1985) p. 5.
17. From the full text (*see* Note 10).
18. Gen.-Lieut. D. A. Volkogonov, 'Fantasty i "zvezdnye voiny" ', *Oktyabr;* **1** (1986) p. 177.
19. V. M. Gavrilov and S. V. Patrushev, 'Gonka vooruzhenii vierarkhii global'nykh problem', *Voprosy filosofii* **5** (1984) p. 99.
20. 'Over the roof of your home: interview with space scientist Boris Raushenbakh', *Twentieth Century and Peace* **6** (1984) p. 28.
21. *See* Pike *op. cit.* (Note 4).
22. e.g. H. A. Bethe, R. L. Garwin, K. Gottfried and H. W. Kendall, 'Space-based ballistic-missile defense', *Scientific American* (October 1984) pp. 37–47; S. D. Drell, P. J. Farley and D. Holloway, *The Reagan Strategic Defense Initiative: a Technical, Political and Arms Control Assessment* (Center for International Security and Arms Control, Stanford University, July 1984).
23. Any edition (*see* Note 7). Elsewhere the cost ratio is given as 'a few per cent' (Ye. Velikhov and A. Kokoshin, 'Yadernoe oruzhie i dilemmy mezhdunarodnoi bezopasnosti', *Mirovaya ekonomika i mezhdunarodnye otnosheniya* **4** (1985) p. 39).
24. Scientists' report (Note 7); Olenev *op cit*. *See also* the discussion in *Detente* **4** (September 1985) pp. 11–12 and **5** (Winter 1986) pp. 18–19.

25. Velikhov, *op. cit.* (Note 12).
26. A term coined by R. S. Ovinnikov, 'Chto stoit za strategiei "zvezdnykh voin" ', *SShA* **11** (1985) p. 15.
27. Scientists' report (Note 7).
28. Ovinnikov, *op. cit.*
29. Scientists' report (Note 7).
30. V. Petrov, 'Za vzaimnuyu sderzhannost'', *Pravda* (28 December 1985) p. 4.
31. D. B. Rivkin, 'What does Moscow think?', *Foreign Policy*, (Summer 1985) pp. 85–105; Meyer *op. cit.* (Note 6).
32. M. Gorbachev, 'Press conference in Geneva', *Soviet News*, (27 November 1985) p. 439.
33. M. Gorbachev, 'Session of the USSR Supreme Soviet', *Soviet News* (27 November 1985) p. 434.
34. A. Vasilyev, 'Programma agressii v kosmose', *Mirovaya ekonomika i mezhdunarodnye otnosheniya* **11** (1985) p. 21.
35. Scientists' report (Note 7).
36. *'Star Wars' . . . op. cit.* (Note 16) p. 42. This concern is elaborated theoretically in terms of an expanding 'field of strategic indeterminacy' by V. A. Gelovani and A. A. Piontkovskii, *SOI i problema balansa strategicheskikh sil SSSR-SShA* (paper presented to the 2nd conference of Soviet Scientists for Peace, Against Nuclear Threat, Moscow (May 1986) mimeo).
37. C. Schmidt-Häuer, *Gorbachev—the Path to Power* (I. B. Tauris & Co. Ltd., London, 1986) p. 160.
38. This is admitted by Velikhov *op. cit.* (Note 12).
39. D. M. Hart, 'Soviet approaches to crisis management: the military dimension', *Survival*, September–October 1984, pp. 214–23.
40. *'Star Wars' . . . op. cit.*, (Note 16) p. 8; Scientists' report (Note 7).
41. e.g. Bethe *et al.*, *op. cit.* (Note 21) p. 46.
42. *'Star Wars' . . . op. cit.*, (Note 16) p. 41.
43. A. A. Vasilyev and S. K. Oznobishchev, 'Militarizatsiya kosmosa', *SShA* **5** (1984) p. 112.
44. *The Night After . . ., op. cit.* (Note 14), pp. 28–30.
45. N. Hawkes, 'Riddle of Moscow's satellites U-turn', *The Observer* (17 February 1985) p. 19.
46. Arbatov *op. cit.* (Note 13).
47. *'Star Wars' . . . op. cit.* (Note 16) p. 27; Scientists' report (Note 7); Olenev *op. cit.* (Note 10).
48. Nekrasov *op. cit.* For a report of an analogos American study, see R. Scheer, 'SDI lasers "will be able to start fire storms" ', *The Guardian* (13 January 1986) p. 8.
49. 'Realizovat' zhenevskie dogovorennosti', *Pravda* (13 December 1985) p. 4.
50. *'Star Wars' . . . op. cit.* (Note 16) pp. 27–34.
51. *'Star Wars' . . . op. cit.* (Note 16) p. 55.
52. Olenev *op. cit.* (Note 10).
53. Volkogonov *op. cit.* (Note 18).

54. *Literaturnaya gazeta* (12 March 1986) p. 8.
55. As Note 31.
56. Nekrasov *op. cit.*
57. For examples of this ideology of space power, *see*: Editors of *Air Force Magazine*, *Space Weapons—a Handbook of Military Astronautics* (Thames & Hudson, London, 1959); and B. J. Smernoff, 'A bold two-track strategy for space: entering the second quarter-century' in F. D. Margiotta and R. Sanders (eds.) *Technology, Strategy and National Security* (National Defense University Press, Washington DC, 1985) p. 139.
58. A. Kokoshin, 'Nuclear poker', *Twentieth Century and Peace* (1986) No. 4 p. 22.
59. P. B. Stares, *Space Weapons and US Strategy: Origins and Development* (Croom Helm, London, 1985), pp. 74–5, 80. *See also* E. P. Thompson (ed.) *Star Wars* (Penguin, Harmondsworth, 1985), pp. 12–3, 130–1.
60. *See* scientists' report (Note 7(c)), p. 76; A. Chapis, 'Danger from space', *Twentieth Century and Peace* (1986) No. 10, p. 12.
61. Capt. Ye. Nikitin, ' "Zvezdnye voiny"—ugroza chelovechestvu', *Krasnaya zvezda* (15 December 1985) p. 3.
62. Chapis *op. cit.*
63. *'Star Wars' op. cit.* (Note 16) p. 9; A. Nikonov, *Twentieth Century and Peace* (1986) No. 3, p. 20.
64. Chapis *op. cit.*
65. Vasilyev nd Oznobishchev *op. cit.* (Note 43).
66. Gorbachev *op. cit.* (Note 32) p. 441.
67. Olenev *op. cit.* (Note 10), full text; 'Mezhdunarodnaya bezopasnost' i voprosy prekrashcheniya gonki vooruzhenii—press-konferentsiya v Moskve', *Pravda* (19 December 1985).
68. Olenev *op. cit.* (Note 10).
69. M. Agurskii, *Sovetskii golem* (Overseas Publications Interchange Ltd., London, 1983) e.g. p. 64. A recent coded attack on the military technocrats took the form of the film *At the Exit of the Night*, reviewed by R. Bykov and K. Lopushanskii, 'Budet li konets sveta?' *Nauka i religiya* **1** (1986) p. 41. 'Today too there are people who demonstrate the impossibility of development without wars, which supposedly facilitate progress in technology. The viewer will meet such "theoreticians" in the film.'
70. Olenev *op. cit.* (Note 10).
71. A. Gromyko and V. Lomeiko, *Novoe myshlenie v yadernyi vek* (Moscow 1984) p. 74. *See also* Drell, Farley and Holloway *op. cit.* (Note 22) p. 26.
72. S. Akhromeyev, 'Utverzhdeniya Vashingtona i deistvitel'nye fakty', *Pravda* (19 October 1985) p. 4.
73. Gorbachev *op. cit.* (Note 32) p. 441. *See also* S. D. Shenfield, 'Soviets may not imitate Star Wars', *Bulletin of the Atomic Scientists* (June–July 1985) pp. 38–9.
74. Speech to Supreme Soviet (*Pravda* 28 November 1985).
75. Gorbachev's speech to Nobel Prizewinners (*Krasnaya zvezda* 14 November 1985).

76. Scientists' report (Note 7).
77. *'Star Wars'* . . . *op. cit.* (Note 16) p. 54.
78. Olenev *op. cit.* (Note 10) full text.
79. Olenev *op. cit.* (Note 10).

8 The Soviets and SDI

Terry McNeill

Until the mid 1960s the predominant view was that there was no possible defence against nuclear attack, especially to a nuclear attack based on long-range ballistic missiles. By 1966, however, the development of radars and related defence technologies had advanced to a point where the Soviets thought it worth while developing an antimissile defence system around Moscow to defend the capital against incoming missiles (the Galosh complex). Faced with the prospect of a reciprocal response on the part of the United States, in both defensive and offensive terms, and sensible of the rudimentary and unreliable state of its own ABM system at that time, the Soviet leadership was eventually won to the view that the ABM systems would not confer the necessary degree of invulnerability. This acknowledgement opened the way to the US–Soviet agreement of mid 1972 whereby further ABM development was limited and both sides, at least nominally, accepted mutual vulnerability based on the threat of an annihilating retaliation as the basis of their security. Reagan's decision to abandon mutual vulnerability as the basis of American security represents the rejection of the strategic doctrine that has been the orthodoxy of the past decade and more.

His case for doing so rests upon two arguments. The first is that the scale of Soviet build-up in all categories of weaponry, but especially in strategic systems, over the 1970s has so upset the force balance between the sides to the point where vulnerability has become asymmetrical. His second argument is that the development of defence technology since the signing of the ABM agreement opens up the possibility of a 'truly effective non-nuclear defence'. Though not made a pivotal argument in the President's statement, a further justification for returning to the search for strategic defence, and one subsequently advanced with increasing assertion by White House circles, is that the Soviets themselves have covertly long been pursuing a similar objective. In the words of the then White House National Security Affairs Adviser, Robert McFarlane, Moscow has in fact 'the most advanced space defence programme on the face of the earth'. Needless to say, the Soviets totally deny this. According to repeated official statements—for instance that of Foreign Ministry spokesman, Vladimir Lomeiko, to Washington pressmen on 28 October 1985, the Soviet Union 'does not have any space weapons' and 'is not engaged in research on space attack weapons'. In reply to such protestations Reagan has bluntly stated that the Soviets refusal to admit that they are engaged in the development of a strategic defence system is 'not only a deception but a dangerous deception'.[1]

SOVIET STRATEGIC DEFENCE

The White House's contention of Soviet superiority in strategic defence is at least as controversial as the arguments relating to the technical feasibility of SDI itself. Any judgement of what the Soviets have been doing in this field relies largely on intelligence sources and upon the ability to attach purposes to ambiguous or multifunctional technologies. It is thus difficult for those not privy to this intelligence, and who lack the technical expertise, to speak in any way other than at second hand. A degree of circumspection is therefore prudent, not only for this reason but also because there must inevitably be some self-serving aspect to US administration statements about what the Soviets possess. And, not least, prudence should be exercised because it is evident that even US government agencies are themselves at odds about what their intelligence data means.

The US case for saying that the Soviets have long been engaged in a sustained, highly ramified and far-advanced programme to create an effective antimissile system has been set out in a number of Defense Department documents and in the statements of Pentagon spokesmen. This gist of what they say is as follows:

ABM Violations

Moscow has contravened the 1972 ABM Treaty; two contraventions are regularly cited. The first relates to the deployment of mobile ABM systems in the Moscow complex (not permitted by the terms of the Treaty). The second, and more serious violation, concerns the development and deployment of a radar system of the ABM type at Krasnoyarsk. Article 6(b) of the 1972 ABM Treaty expressly prohibits either party to the agreement deploying radars for early warning of stategic ballistic missile attack, except at locations along the periphery of its national territory, and enjoins that they be oriented outwards. Krasnoyarsk violates this stipulation in two ways. Firstly, the radar system is not located within a 150-kilometre radius of the national capital, as required of ABM radars, nor is it on the periphery of the Soviet Union and pointed outwards, as required for early-warning radars. Secondly, the radar system is, it is believed, for the purpose of ballistic detection and tracking, including ballistic missile early warning, and not for space tracking, as the Soviets proclaim, a role that would be permissible under treaty provisions. In fact Krasnoyarsk is seen as being a part of a network of large phased-array radars which integrate into an arc of coverage from the Kola Peninsula in the north-west USSR, around Siberia, to the Caucasus in the south-west. The Krasnoyarsk element of this network is the only part that is viewed as a treaty violation.

Furthermore, it is also alleged that the USSR has over recent years been developing several types of rapidly deployable radar systems of an ABM type. One type depends upon a small transportable radar (the flat twin engagement radar) which, it is believed, could be teamed with interceptor missiles to make up a mobile, easy to erect territorial ABM defence system. Another possible way of achieving such a system, it is foreseen, is through the use of a modified surface-to-air missile (SAM) system. SAM missiles have been tested for such purposes in the past; the expectation is that a soon-to-be-deployed new SAM missile, the SA-X-12, will be specifically an ABM missile, albeit in another guise. These developments would all represent treaty violations of major proportions.[2]

The second main contention is that the Soviet Union is engaged in the development of a number of advanced technologies for defence against ballistic missiles. These include:

Lasers

The Soviet laser weapons programme is held to involve over 10,000 scientists and more than half a dozen major research and development facilities and test ranges. The Sary Shagan Missile Test Centre where the key work goes on is said to have facilities for several air defence lasers, a laser that may be capable of damaging satellites in orbit and a laser that could be used in feasibility testing for ballistic missile defence applications. Costed in US dollars, the Soviet laser programme is estimated to be in the range of roughly $1 billion per year.

The Soviets are also said to be conducting research into different types of gas lasers, with likely weapons applications; into gas dynamic lasers; into electric discharge lasers; into chemical lasers; into X-ray lasers; and other similar front-edge technologies, some of which, like the rocket-driven magnetohydrodynamic generator, have no counterpart in the West. They are also said to be carrying out advanced work on developing the optics necessary to the development of a space-based laser system.

Unlike the US, the USSR, according to the Defence Department, has advanced in some areas beyond research and is already possessed of ground-based lasers that could be used to jeopardise US satellites and which provide the prototypes for space-based antisatellite laser weapons which could be available for deployment within the decade.

Particle Beams

According to the US administration, the Soviets have made 'impressive' strides towards acquiring a space-based weapons capability using particle beams against satellites and in general to be well ahead of the US in understanding how to use particle beams for weapon purposes.

Kinetic Energy

The field of kinetic energy weapons is also one where the USSR is held to have made significant advances, to a point where it is believed they could deploy a close-in attack, space-based system employing manoeuvring satellites in the near future, though it is acknowledged that a long-range system is beyond them until the mid 1990s or even later.

ASAT

In antisatellite development the USSR is also held to be ahead, having the world's only operational system—a co-orbital device that homes in on the target satellite and destroys it by exploding a conventional warhead.

Air Defence Systems

Accompanying the exotic technologies, the USSR continues to invest enormous resources in a wide array of strategic and tactical air defence systems—interceptor aircraft, airborne warning and control systems (AWACS), over-the-horizon and air surveillance radars, a SAM capability that provides a low-to-high-altitude barrier, area and terminal defences under all weather conditions. It is also investing heavily in other forms of passive and active defence. Taken together, these capabilities, in the eyes of the US State Department, seriously erode the retailatory capabilities of US strategic forces upon which deterrence rests. America's SDI is therefore needed, it is argued, to counter a trend that is strongly in the Soviets' favour and as a necessary insurance lest the USSR attempts to capitalise on its advances to install a comprehensive antiballistic missile system in the near future.

This is in outline the American case. Some of the particulars of the argument have been strongly contested. The role of the Krasnoyarsk radar is, according to the sceptics, not nearly as clear cut as US officials make out. External observation, they say, is not enough to determine its function. They argue that without on-site inspection that includes an examination of the software on the computers as well as the equipment itself there can be no certainty in the conclusion that it is a battle-management system and not what the Soviets say. Others contend that even if it is true that Krasnoyarsk represents a violation of the ABM Treaty, it is a minimal infringement compared with the violation to this treaty, not to mention the Outer Space Treaty, implied by the pursuit of SDI—a view that is little softened by the US's re-interpretation of the ABM treaty to permit testing for space-based defences.

Still others focus on the ASAT argument, saying that in this respect too the case against the Soviets has been greatly overstated. While acknowledging that it is true that for at least a decade longer than the United States the USSR has had an antisatellite weapon, based on the homing-in and self-destruct principle, they point out that this technology is rudimentary in comparison with the antisatellite capability being developed by the United States, which includes the use of rockets fired directly at the target from an F-15 fighter and a metal net system deployed by satellite.

It may be said too that there is considerable scepticism about the idea that the Soviets have a lead in space technology overall, which seems to have gained increasing currency in the wake of the Challenger disaster. Jane's Space Flight Directory, for instance, claims that 'the Soviets are so far ahead of the United States in space that they are almost out of sight'. It is true that the Soviet Union conducts more launches than the United States (ninety-eight in 1985 against seventeen for the United States). But, as Paul Stares of the Brookings Institute argues in a recent press comment, frequent Soviet launches may be explained by the much shorter lifetimes and lower reliability of their satellites, many of which fail prematurely. He also points out that the USSR is still employing the family of boosters that was first used to launch Sputnik in 1957, hardly an indication of an innovative and flourishing space programme.[3]

As to Soviet laser technology, many argue that what the Soviets are engaged in is basic research and not an active weapons programme. Even the Pentagon documents talk only about 'possible' military applications. Indeed, those who tend to pooh-pooh claims of a major Soviet lead in the lasers field can cite authoritative testimony to the contrary made before the US Congress. This testimony includes a statement by the US Air Force in the spring of 1985 to the House Appropriations Subcommittee to the effect that, despite years of research on lasers and particle beams, which someday, it is conceded, may be used to shoot incoming warheads, the Soviet Union 'has no identifiable lead in the application of these technologies to a space-based defense'.[4]

Similarly, a CIA assessment made public by the Massachusetts Representative, Edward J. Markey, in May 1986, again in contradiction to testimony before Congress by Reagan Administration officials, claims that the Soviet Union would have to conduct new underground nuclear tests before it could develop and deploy an X-ray laser. In fact, some American scientists have estimated that perfecting such a laser system, which focuses the power of hydrogen detonation beams of radiation, might require dozens, even hundreds, of underground nuclear tests.[5]

It would seem, therefore, that as long as the Soviet Union continues to adhere to its declared moratorium on nuclear testing it follows that little progress towards the completion of this technology can take place. In parenthesis, as it were, it may be added that since this assessment was made the Soviets have suffered a serious blow to their laser programme, following the loss of their only airborne laser weapons laboratory—a converted Ilyushin IL-76—which, reportedly, blew up at the Schelkino air base in mid 1976.[6] The US Defense Department's claim of any significant Soviet lead in SDI technology seems to be seriously undercut by its own further admission that in 1985 the USSR was behind in fifteen fields of technology and ahead in none.[7]

In his 1985 annual statement to Congress the Undersecretary of Defense for Research, Donald H. Hicks, noted that the Soviet Union 'does not surpass the United States in any of the 20 basic technologies that have the greatest potential for significantly improving military capabilities in the next 10 or 20 years'.[8] Lieutenant General James A. Abrahamson, who heads the SDI research programme, and John K. Sellers, chief of the Defense Intelligence Agency's strategic defence branch, have likewise stated that the USSR lags behind America in the key technology of computers needed for battle management of a space defence system, though in optical computers they have a strong programme which has the potential to give them the advantage for some

SDI-related tasks. Also, their assessment of the respective sides development of radar technologies indicates a fairly balanced picture of strengths and weaknesses.

Other expert testimony, while not denying the administration's claim of the scale of the Soviet strategic defence research effort, tends to give a more restrained projection of the likelihood of it becoming anything very serious to reckon with in the near future. The consensus is that the greater part of current Soviet strategic defence programmes appears to be directed towards traditional, less exotic purposes, primarily 'land-based radars that direct land-based launchers capable of firing a missile at an incoming warhead'.

Some of the inconsistencies in US administration statements on Soviet strategic defence, apart from reflecting an apparently irredeemable tendency to air interagency wrangles in public, seem to stem from the fact that the US administration is trying to respond to contrary pressures. On the one hand, the administration has sought to impress the urgency of SDI for the United States by dramatising a Soviet lead in various technologies (shades of Kennedy and the missile gap?). On the other hand, there is the need to convince a sceptical Congress that, despite what the detractors say, SDI, as Reagan has pictured it, is technically feasible and warrants every last buck demanded. This gives rise to a tendency to magnify US advances, talk about 'breakthroughs' in the field and optimise projections of further progress.

But, irrespective of the doubts aroused by inconsistencies between different versions of the administration's case, and the occasional hyperbole that has accompanied it, what gives plausibility to the argument that the Soviets have been seriously pursuing some kind of strategic defence programme, and may well be ahead in certain aspects of it, is that such a programme makes perfect sense, whether one takes a dovish or hawkish view of Soviet intentions. For instance, it is not necessary to subscribe to the existence of some Kremlin masterplan for global dominance to acknowledge that it is a long-standing Soviet aim to alter the East–West 'correlation of forces' in its favour.

In Soviet parlance the term 'correlation of forces' has a complex meaning which includes economic, political and sociological elements, as well as purely military considerations. But as far as military considerations go, the crucial one is, of course, the US–Soviet military balance. In relation to this, the USSR has long been consistent in its objectives. Firstly it seeks to redress the strategic imbalance that in the 1950s and early 1960s existed between themselves and the Americans, and which was so embarrassingly displayed to the world by the Khrushchev/Kennedy confrontation at the time of the Cuban missiles crisis. And secondly, having achieved that, it seeks to maximise American vulnerability while minimising its own. The logic of this is that it has never made any sense to Moscow that she should accept deterrence in the sense that the West has embraced it, that is equal vulnerability involving mutual assured destruction. All along she has striven to attain security on the basis of military superiority and territorial invulnerability. Hence the 'we build and they build; we stop and they build' pattern; hence the continuing commitment to ABM and so on. It is just as obvious to Moscow that she should pursue the goal of acquiring an impregnable strategic *defence* capability as she should acquire a preponderant strategic *offence* capability.[9]

Soviet leaders are on record in assuming that any major East–West military conflict is likely to go nuclear; this has been repeatedly stated in their military writings. Indeed, the Soviet former Chief of Staff, Marshal Ogarkov, in response to Western conjecturings about the possibility of a 'limited nuclear war' (i.e. one confined to Europe) specifically warned that such scenarios were dangerous fantasies. As far as the USSR was concerned, he warned, a major European engagement between Nato and the Warsaw Pact would not only be nuclear, but it would inevitably be *strategically* nuclear (i.e. intercontinental).

In the event of nuclear war occurring, it is the clear Soviet intent to prevail, if possible, in such a conflict, while minimising damage to its own assets. This in itself is neither surprising nor particularly reprehensible ('better red than dead' obviously means something different viewed from the Kremlin than from the White House). Besides, the United States too has contemplated the possibility of having to fight a successful nuclear war. As long ago as 1962 Defense Secretary Robert

McNamara declared 'we will use nuclear weapons to prevail, if this becomes necessary'. His later successor James Schlesinger said something similar; President Carter in turn stressed the need for a survivable command and control system in the event of nuclear conflict. The difference between Reagan and his predecessors is that under him the US has begun to spend the money to acquire the means whereby this may be possible.

To this end the US administration has drawn up and begun to implement a $40 billion plan for the creation of a military infrastructure (C^3, means for rapidly reprogramming nuclear missiles under conditions of war, alternative communication and command facilities and so on).[10] In saying these things and in adopting such measures the USA is doing no more than taking a leaf out of the Soviets' book.

Soviet military doctrine specifically enjoins that all possible measures should be taken to ensure the wartime survival and functional ability of Soviet nuclear forces, civilian and military commands, communications, key industries, local leaderships and as much of the general population as possible. US intelligence sources have estimated that the USSR spends at least 50 per cent of its budget for strategic forces for these purposes. This has led to a staggeringly impressive array of both active and passive measures for national defence. On the passive side, there is at present provision for housing in hardened shelters away from the likely sites of missile strikes for some 200,000 key party and government personnel; reserves of vital materials are maintained; extensive blast shelters for the industrial workforce have been built; and there is a steady programme of hardening ICBM silos and launch facilities. On the active side, in addition to the acknowledged ABM facilities around Moscow, which have been upgraded and strengthened over the years (while the USA has abandoned theirs), and the other similar measures discussed above, the USSR has an air defence system of some 10,000 surface-to-air missiles and thousands of interceptor aircraft.[11]

All this is evidence on the Soviet's part of a determination to be prepared for, to prosecute if necessary, and survive if possible, a protracted nuclear engagement. However, neither Soviet preparations nor the more recent US efforts should be viewed as a contradiction of the statement made by Reagan and endorsed in similar sentiments by Gorbachev: namely, that nuclear war 'can never be won and must never be fought'. What it means is that both sides, not just the Soviets, now subscribe to a doctrine of 'damage limitation'. The leakproof astrodome envisaged in SDI, of course, were it to be realised by either side, would invest 'damage limitation' with a qualitatively different character.

SOVIET ANXIETIES ABOUT SDI

It is not always easy to separate propagandist posturing from genuine anxieties in Soviet reactions. The campaign against the so-called neutron bomb in the late 1970s and the subsequent highly orchestrated efforts to frustrate Cruise and Pershing II deployments smell more of propaganda than of real Soviet fears of pre-emptive or first-strike intentions on the part of NATO. Whilst the present wholesale attack on SDI undoubtedly also has a propaganda purpose, there is some reason to believe that the Soviets this time do have genuine worries. That is not to say that as yet they are any more convinced than Western critics that the initiative will work, at least not in the visionary form in which it was first unveiled. But, in so far as the gush of pronouncements about SDI properly reflect official thinking, they seem to indicate considerable anxiety, perplexity and uncertainty. The present utterances lack the usual consistency that one normally associates with the well-oiled propaganda campaign. Statements veer from one extreme to the other. One day the line is that SDI is useless, expensive and easily neutralised by counter-measures. The next day they denounce it as in reality a 'strategic offensive initiative' which is deeply threatening, destabilising and provocative.[12]

One can understand why it gives rise to this ambivalence. SDI is a step into the unknown which takes the arms race into a high technological terrain where the old familiar yardsticks of quantitative balances are no longer reliable guides. The Soviets' worst nightmare must surely be that the US will achieve such a level of strategic defence that will consign the Soviets' expensively acquired arsenals of ICBMs to the scrap heap. Even if they are to adopt the more sanguine assumption that SDI will be only a partial success, it still must vastly compound the problem of their nuclear planners by making it impossible for them to forecast just how many of their warheads would break through or precisely which targets they would hit (though if city busting were substituted for counterforce targeting the problem would admittedly lessen also; as a Rand Corporation team has observed, an American SDI increases the rationale for the Soviets striking first). In its very least effective form, SDI must be seen as providing the US with the potential for some kind of point defence system, thereby investing the Minutemen sites with a high degree of immunity to Soviet strikes.

Even if SDI were to be a total flop as far as attaining any level of strategic defence, this is hardly something that the Kremlin dare gamble on nor necessarily draw much comfort from. The Moon shot, the Manhattan project and numerous other American firsts are a salutory reminder that today's technological chimera can soon become tomorrow's technological commonplace. But this apart, there are the probable collateral gains which will spin off from a project that is funded to the extent projected for SDI and which harnesses the best scientific brains of the Western world; these could weigh heavily in the military balance. Here one thinks of the gains in computer development, radars and guidance technology, in identification and weapon targeting, in lasers, optics, new forms of firepower and so on, all of which will give an enormous boost to conventional war-fighting capacity even if SDI as such were to wither on the vine. Any way one reads the tea leaves, the future is set for a new type of arms race at a much higher technological level.

It is not difficult to understand why the Soviets see in any version of SDI a major military challenge, an attempt on the part of the United States to occupy the military high ground and, given their neuroses about security, a possible bid to achieve a first-strike capability (an argument advanced by Kremlin US specialist Arbatov, among many others). As to the latter, if the Soviets cannot achieve such a position for themselves, they are absolutely adamant that no one else will achieve it.[13]

ARMS CONTROL TALKS

The recent rush by the Kremlin to trade concessions in various categories of arms and to sacrifice positions such as on-site verification, in Mutual and Balanced Force Reduction talks etc., from which they were previously immovable is not just a reflection of Gorbachev's pragmatism, as some argue, though he is relatively pragmatic—one has to give him the credit for being the first Soviet leader who apparently appreciates that security is not a unilateral state—it is also impelled by an urgent desire to strangle SDI at birth.[14]

Apart from purely military considerations, there are also pressing domestic reasons why Gorbachev wishes to avoid a new cycle in the arms race. The serious plight of the Soviet economy, now made more acute by the sharp decline in hard-currency earnings, is an open secret. Numerous Western experts (Alec Nove, Stephen F. Cohen, Jerry Hough, Seweryn Bialer, Philip Hanson, to mention but a handful) are all agreed that restoring economic growth is the General Secretary's number one priority. So far he is seemingly bent on achieving this without taking the risky road to marketisation but rather through shifting resources into the sluggish industrial sectors, increasing incentives for productivity, a bit of whip cracking and by streamlining the managment process.

Cramped for resources, the diversion of further funds into meeting the challenges of SDI inevitably puts his whole strategy for recovery at risk by threatening to syphon so much of the budget into military procurement (presently reckoned by US intelligence agencies to absorb about 15–17 per cent of GNP) as to condemn the economy to unstoppable decline.

Gorbachev wants an arms agreement which will cut off significant areas of arms competition with the West and, in particular, he wants to stop SDI. In so far as these objectives run foul of internal opposition, he is better placed to get his way than any leader probably since Stalin. All the evidence shows that he is a determined and resourceful man and determined too to impose his will upon the military and resist pressure for more money. To this end, he has taken steps to weaken the military voice in top policy-making circles: the septuagenarian Minister of Defence, Marshal Sokolov, unlike

Ustinov and Grechko before him, has been denied entrance to the inner power elite; the strong-man, Ogarkov, remains on the sidelines; General Yepishev, who expressed misgivings about arms talks, has been dropped, as was the extremely hard-line Romanov. At the same time, a spate of new appointments to command and command-related positions has greatly enhanced the Party leader's leverage over the men in uniform. The people most strongly opposed to arms limitation have been eclipsed; the people who have taken their place and are his key advisers, such as Dobrynin and Yakovlev, are more amenable to talks and, incidentally, much better acquainted with the West than those they replace.

Also he has made a deliberate move to reduce the influence of the High Command in defining military priorities. Steven Meyer of the Massachusetts Institute of Technology has noted that the recent Soviet Communist Party Congress (Feb./March 1986) approved a change in the description of how military spending levels would be determined so as to permit political leaders to lower the levels. The new standard, he says, is to fund the military so it can prevent 'strategic superiority of the forces of the imperialists' whereas the previous formulation was a pledge to provide the Soviet military with whatever it believed it needed 'to reliably defend the homeland'. Under the old formulation the Soviet Chief of Staff would have the final word on what was needed. Now the Party leadership will be able to insert its views on what the United States is doing as part of the mix.[15]

But as far as cutting arms deals are concerned 'it takes two to tango' and the present indications are that the Reagan Administration is unable to accept Soviet terms, especially the Soviet proposal for a moratorium on space defence in return for deep cuts in strategic arms.

THE SPECIALISTS' DEBATE

The White House's commitment to SDI, it has been repeatedly stated,is not negotiable, however many SS-18s Gorbachev is willing to throw on the table. Indeed there are those in the administration who argue that no deal with Moscow is worth while, whatever the inducements, in the light of Soviet bad faith in upholding existing treaties. Some in fact believe that, instead of seeking arms reductions, now is the right moment to up-the-ante on the Soviets, their argument being that the West can and should force radical change upon the Soviet Union by intensifying arms competition to a level that the enfeebled Soviet economy cannot sustain. Hence they advocate an embargo on all kinds of economic interchange with the USSR that might ease her difficulties and urge the White House to speed up military competition at all levels, particularly in the areas of high technology represented by SDI, on the assumption that ultimately this will make the Kremlin crack. In taking this position they have at their back such influential critics of the USSR as leading Soviet specialist, Richard Pipes, and former Soviet dissident, Andrei Amalrik, both of whom have argued that it will take a major crisis such as would be brought about by a collapse of the Soviet economy to bring about real change in the system.[16]

The assumption of those who urge a negative stance on arms control is that the Soviet Union has only a limited capacity to meet the new American challenge because, like Kenneth Adelman, head of the US Arms Control and Disarmament Agency, they belive that in arms production 'the Soviets have their accelerator near or on the floor'. They are reinforced in their view by the findings of a recent CIA-sponsored review of the Soviet economy chaired by Henry S. Rowen, a senior research fellow at the Hoover Institute on War, Revolution and Peace, which picture the Soviet economy as facing 'a terrible situation' with little ability to finance a major weapons build-up. In passing, it may be noted that the CIA and the Pentagon have been at odds over the data for Soviet arms build-up in the last decade, with the Pentagon saying that Soviet procurement of major weapons systems has been running at an annual figure of 5–10 per cent in the early 1980s, whereas the CIA put it at about 2 per cent, though more recent figuring has narrowed the gap between them.

Other specialists claim that the argument that the Soviets are at the limit of their arms output capacity and can be pushed off the map is faulty for, among other things, it overstates the seriousness of the USSR's economic crisis while underestimating the ability of the Soviet elite to mobilise the people in a rally-round-the-flag patriotic response. For instance, Dmitri K. Simes, a leading Kremlin watcher, says it is a mistake to see Russia as on the ropes; in Gorbachev they have a vigorous and able leader with a self confidence bordering on arrogance. A leading specialist on the Soviet military, David Holloway, feels that though Soviet economic problems are bad they are not so bad that they cannot stay in the arms race. If the aforementioned Pentagon consultant and MIT fellow Stephen M. Meyer is right, the Soviets certainly still seem to have the capacity to shift their military production into a higher gear. According to his data, between 1980 and 1984 they built more than 800 new intercontinental ballistic missiles, whereas the USA produced none. He believes, too, that they have about 1,000 missile boosters or rockets stored but not deployed.[17] If, therefore they were to meet the challenge of SDI by missile saturation in the manner discussed elsewhere in this volume by Stephen Shenfield, then the capacity to do it may be there without imposing any excessive new budgetary strains.

Besides, it has been argued that, whatever the truth is about Soviet military capacity, the most likely outcome of a 'squeeze them till the pips squeak' approach is that it will be the opposite to that intended: namely, it will stiffen Soviet resolve and harden their attitude. Stephen F. Cohen would go further, arguing that it is likely to undercut Gorbachev, put paid to a historic opportunity to end the arms race and bring to the fore the hawks in the Kremlin,[18] a view evidently endorsed also by veteran Soviet diplomat and Kremlin watcher, George Kennan. To which it might be replied: not only is it an old chestnut of faulty Kremlinology to see the incumbent leader as a dove at the mercy of hawks (as Stalin was often pictured by witless Westerners), but is it not the case that Reagan's tough stance has wrung more out of the Soviets that all the conciliatory postures of his predecessors?

CONCLUSION

The age of space militarisation has arrived and it is probably too late to prevent it, though the pace at which it happens can possibly be altered. The SDI venture to which President Reagan is committed is but the most ambitious and systematic version of a series of military space programmes which both superpowers have been pursuing for a number of years. Even if Gorbachev were successful in agreeing a deal with Reagan whereby SDI as such were abandoned in return for a package of cuts in offensive arsenals, the long-term impact of this on space militarisation is not likely to be significant. Firstly, it cannot be too often repeated that the arms race is the result not the cause of tension between East and West and as long as that tension persists security will be sought in military terms and, inevitably, in terms of the most effective technologies available. The technology generating the vision of strategic defence and the technology being generated by its pursuit promises to raise military effectiveness to qualitatively new levels. Which power is going to forgo the best for the sake of a piece of paper? To do so would be an act of greater trust than has yet been shown by either side. Secondly, even if SDI *per se* were outlawed, it is likely to continue anyway, albeit in another guise. The flaw in the existing ABM Treaty is that it leaves room for the technologies that are at the heart of a strategic defence system. As Gerard Smith, the architect of the 1972 ABM Treaty admits. 'The great loophole in the ABM Treaty is not whether it permits the development of new exotic technologies, but whether under the guise of anti-satellite weapons and anti-tactical ballistic missile systems and radars, anti-ballistic missile defenses will emerge anyway'. Anyway, treaty or not, these technologies are but the next reach of military science.

Probably the best that Gorbachev can hope for in any negotiation is to slow the American effort in the hope that this will allow the Soviets to stay in the race, and for this he seems willing to countenance deep cuts in the USSR's strategic arsenals and perhaps to trade other things too.

Assuming those who expect the Soviet economy to cave in soon are wrong, then experience suggests that the Soviets will eventually match whatever it is the US comes up with—by copying, by stealing, by sustaining the kind of commitment to defence impossible in an open society and by employing counter-measures. This being so, the likelihood is that the foreseeable future will see a mixture of offensive and defensive systems on both sides but no leak-proof astrodome, no technical fix for what is at bottom a political problem: in short more MADness rather than less.

NOTES

1. Quotation from a Reagan speech in the *International Herald Tribune* (15 October 1985). For a sample of Western critics' response to the Reagan initiative, *see* John Tirman (ed.), *A Space-based Missile Defense: The Fallacy of Star Wars* (Vintage, New York, 1984) (a compendium of the views of the Union of Concerned Scientists); Frank Barnaby, *What on Earth is Star wars? A Guide to the Strategic Defence Initiative* (Fourth Estate, London, 1986) and E. P. Thompson (ed.), *Star Wars: Science-Fiction, Fantasy or Serious Probability?* (Penguin, Harmondsworth, 1985).
2. For an authoritative and detailed account of radar technology and a presentation of the case against the Soviet Union, *see* Frank G. Gaffney, 'A comparison of US and Soviet perspectives on strategic defense', CERIS, Paris (18 October 1985). Mr Gaffney is US Deputy Assistant Secretary of Defense for Nuclear Forces and Arms Control Policy. For a more comprehensive statement on Soviet capabilities, *see also* the US Department of Defense's, 'Soviet strategic defense programs'.
3. For a longer treatment of space rivalries *see* Paul B. Stares, *Space Weapons and US Strategy: Origins and Development* (Croom Helm, London, 1985).
4. *International Herald Tribune* (1 October 1985).
5. *International Herald Tribune* (30 May 1986).
6. *Guardian* (21 August 1986).
7. *International Herald Tribune* (14 February 1986).
8. *International Herald Tribune* (1 October 1985).
9. There is a considerable body of literature, both Soviet and Western, on the concept of the 'correlation of forces'. A useful summary of it in English is provided by David Holloway, *The Soviet Union and the Arms Race* (Yale University Press, New Haven, 1983) pp. 82–3.

   For an analysis similar to the author's of the Soviet attitude to mutual assured destruction, *see* Alun Chalfont, *Star Wars, Suicide or Survival?* (Weidenfeld and Nicolson, London, 1985) Chs. 4 and 7.
10. For details of this programme, *see International Herald Tribune* (29 July 1986).
11. For an outline of these preparations *see* State Department documents cited above (fn. 2) and also Chalfont *op. cit.*
12. *See* Chalfont *op. cit.* pp. 93–4 on contradictions in the Soviet position.
13. General Chervov, a senior spokesman for the USSR Ministry of Defence, has repeatedly laboured the point that SDI is intended to confer

on the US first-strike capacity: 'The SDI is the US drive to achieve decisive military superiority through space . . . the space strike weapons together with the strategic offensive weapons increase an offensive potential and give it—the offensive potential, the strategic offensive weapons—the capacity for an unpunished first strike with the purpose of maximally weakening the replying strike . . . in a word, not a shield but a space sword is being created and that is the military essence of SDI'. Drawn from a Soviet television interview (3 February 1986) as monitored by *BBC SWB Monitoring Report*, Part 1 (6 February 1986).

Chervov's view was given the official imprimatur by a *Pravda* editorial (30 January 1986) which stated: 'The "Star Wars" programme incorporates the claims of earlier years to the attainment of US military superiority. Militarist circles are unwilling to reconcile themselves to the idea of strategic parity which has been achieved by socialism. They still entertain the illusion that this parity is temporary and that by implementing the "Star Wars" programme they will be able to turn the clock back. They desire to use space to achieve decisive military superiority and create the potential and launching a first nuclear strike with impunity—that is the true purpose of the notorious SDI.'

14. Gorbachev restated his acceptance of security as a mutual state in his report to the Twenty-seventh Party Congress. This point was subsequently taken up by Aleksandr Bovin, a leading *Izvestiya* commentator, who also doubles as a front-line interpreter of the Kremlin to the West. Predictably, in one of his columns (23 April 1986) he underlined Gorbachev's point: 'In our time real security cannot be a result of technical advances, still less unilateral advances. Real security presupposes a political accord aiming to reduce and then eliminate altogether the existing nuclear missile potentials. This is the meaning of Gorbachev's proposals made public on January 15th.'
15. *International Herald Tribune* (24 June 1986).
16. For a succinct statement of Pipes' arguments, *see* R. Pipes, 'Can the Soviet Union reform?', *Foreign Affairs* **63**, No. 1 (Fall 1984). Similar arguments are advanced with great cogency and force by several of the contributers to A. Wildavsky (ed.), *Beyond Containment: Alternative American Policies Toward the Soviet Union* (ICS Press, San Francisco, 1983) and by A. Shtromas, *To Fight Communism: Why and How* (Professors World Peace Academy, New York, 1985). *See also* Andrei Amalrik, 'The Soviet Union—approaching 1984', in R. Wesson (ed.), *The Soviet Union: Looking to the 1980s* (Hoover Institution Press, Stanford, California, 1980). Some of the arguments central to this debate are advanced also by Jeane Kirkpatrick, 'American foreign policy in a cold climate: a long conversation', an interview between the author and George Urban in *Encounter* **LXI**, No. 3 (November 1983).
17. For details, *see International Herald Tribune* (24 June and 26 June 1986) and *ibid.* (20 December 1985).
18. Stephen F. Cohen, 'Gorbachev's detente: dangerous illusions', *International Herald Tribune* (3 June 1986).

Part III
Implications

9 The Military Uses of Space and European Security

Stephen Kirby

As an actor Ronald Reagan may or may not have known his Shakespeare but as President he has avoided the dilemma of the Prince of Denmark. Hamlet declared that he could count himself '. . . a king of infinite space, were it not that I have bad dreams', but the President's regal and 'theatrical dream'[1] for space leaves his somnolence untroubled. Others, his European allies prominent among them, sleep less well. Their concern is that the President has not simply accelerated the transition from a 'passive' to an 'active' use of space with the promotion of strategic defences and related systems: it is more that these technologies will revolutionise strategic policy, superpower relationships and coalition politics. Yet, the President's major programme, the SDI, lacks 'an overall concept definition'[2] and his administration has yet to show that it has considered its collateral effects in any depth.

THE PACE AND SCOPE OF CHANGE

Space has long been militarised since the great majority of satellites have always served military purposes, and their importance for the warning, conduct and termination of terrestrial war, both nuclear and conventional, is increasing. The deployment of systems like NAVSTAR and its Soviet counterpart GLONASS substantially enhance the accuracy of strategic offensive forces but also provide tactical and theatre commanders with data in real time which can be vital for conventional war fighting. This latter development has already promoted studies of new opportunities for, and threats to, the conventional force structures of the Western Alliance. However, space will soon be 'weaponised' as the SDI and Soviet developments and responses fuel superpower plans for strategic missile defences and also drive forward technologies for ASAT attack, satellite defence (DSAT), and for ATBMs.

The managers of the SDI programme anticipate incremental and sequential deployment of some strategic defensive technologies by the early 1990s with extensive deployments after the year 2000.[3] Within the same period the deployment of effective ASAT systems is also possible and likely, not least because there is considerable overlap between the SDI's strategic defence technologies and those required for the destruction of satellites. The Soviet Union already possesses a crude co-orbital ASAT system and the United States has recently tested a much more capable system based upon an air-launched miniature homing device. Even though its testing has been temporarily halted by Congress, the ban is under pressure[4] and the prospect increases of an early deployment of an American ASAT and the improvement of the existing Soviet system. In the attempt to win allied support for the SDI, the development and deployment of an ATBM has been forced onto the NATO agenda. This has occurred in part because European governments are seeking an alternative to SDI participation more in keeping with their regional defence interests but also because SDI research is more likely to provide effective technologies for terminal than for strategic defence.

Technological limitations and even new arms control agreements may yet deny the full realisation of the President's dream of effective strategic defences, but SDI has made a comprehensive and vigorous superpower weapons competition in space—long feared by the European members of NATO—an imminent reality.

SPACE WEAPON DEPLOYMENT

Despite the equivocal attitude of European governments to the SDI, the consideration of its strategic, political and economic implications cannot be deferred. And yet the Western allies have persisted in a discreet fiction of describing SDI as a research programme that will provide a future President and Congress with the technical information necessary to decide the feasibility of development and deployment in the early 1990s.[5] The kernel of the fiction resides in the suggestion that there will be a distinct point in the future at which America, and perhaps its allies, can decide whether or not to deploy an SDI system. In Europe the fiction, or more accurately the illusion, consists of the suggestion that there is still time before governments need confront the many problems associated with the deployment of strategic defences and the ASAT and ATBM technologies they will promote. It is true that the report of the Strategic Defence Study Team (the Fletcher Panel) supported these fictions by promoting a plan, now administered by the SDIO, which emphasised technical research and experiments for the demonstration of 'total systems effectiveness in ballistic missile defence *rather than an early deployment capability*.' But there is already a ponderous inevitability about the deployment of many programmes that fall under the SDI umbrella: the SDI seems now to have acquired an 'unstoppable momentum' even though its implementation may be partial and incremental.

The recommendations of the Future Security Strategy Study Team (the Hoffman Panel) more accurately indicate the likely deployment policy for SDI. The panel argued strongly for strengthening retaliatory deterrence by deploying partial defence systems based upon more modest goals than the presidential vision of perfect defences. The Study Team recommended that this could begin almost immediately with the use of available technology for an ATBM based in Europe, and observed that this would complement plans for incremental and sequential deployments of defensive systems in the 1990s. Nevertheless, the SDIO recognises that the Reagan Administration has given guarantees to European governments that deployment will be a matter for allied negotiation.

Those guarantees may now count for very little. The President announced on 12 March 1986 that he might consider the partial deployment of the SDI before it had been fully developed thereby reversing, without allied consultation, earlier statements made in March and November 1985. More recently Casper Weinberger stated that nothing should 'hamper or delay in any way our ability to *deploy* a strategic defense system' once research had proved the system to be feasible. This policy statement was also made without consulting allies, despite its flat contradiction of the President's undertaking to Mrs Thatcher at Camp David in December 1984.[6] These statements, together with the industrial and Congressional muscle already mobilised behind the programme, make early and partial deployment of SDI more likely than not.[7] It is this that has made SDI projects and their closely related programmes for ASAT and ATBM not merely durable features of US defence policy, but ineradicable elements of US defence funding and planning.

Deployment of these systems, however, is simply a future complication of the central problem that already faces the Europeans, and that is an American President utterly committed to the energetic pursuit of strategic defences and to the recasting of a strategic philosophy of twenty years standing. These facts alone, irrespective of the future feasibility of strategic defence, have changed the perceptions and are progressively changing the responses of allies and adversaries alike to the USA. The strategic verities of the last twenty years have been challenged and cannot be fully restored; for the Europeans it remains only to assess the implications of the changes already under way and to begin the task of identifying the parameters of a new and workable security regime. To do that it is necessary to understand both the extent to which the United States and Europe currently share security interests in space, and the nature of European concerns about the rapidly diverging perceptions of those interests on either side of the Atlantic.

SHARED SECURITY INTERESTS IN SPACE

At present the Western Alliance deploys only satellites in space and these are usually regarded as passive, although their effect is far more ambiguous than this description may suggest.

Satellites have always supported both the mechanisms that underwrite interbloc stability—early warning, arms control verification and crisis management—and the capabilities that enhance war-fighting potential—planning, guidance and battle management. There are currently three kinds of shared interest in space among the members of the Western Alliance. The first is the co-operative arrangements for the deployment, tracking and operation of satellites, the second is the sharing of the facilities and information that satellites provide and the third is the importance of space assets for the effectiveness of allied defence and deterrent policies.

Although the overwhelming proportion of Western military satellites are owned by the United States they depend, in part, for their successful operation on ground stations located throughout the Alliance. For example, the Morwenstow base in Cornwall handles the signals intelligence from the CIA's RHOYLITE satellites, and the British Skynet ground station at Oakhanger is used for US satellite control.[8] The operation of a US ASAT system would also depend upon facilities like the phased-array radar at Fylingdales and ground-based electro-optical deep space surveillance (GEODSS) systems sited, among other places, in Portugal. NORAD uses information from these and other ground stations for space surveillance, satellite protection and will be able to use it for 'satellite negation'—the official term for ASAT operations.[9] Until recently European military satellites were either leased or purchased from the United States, as was Britain's Skynet 1, and placed in orbit by US launchers, as was the British-built Skynet 2B and the NATO tactical satellite communications programme, which links the governments of NATO countries and is fully compatible with the US Defense Satellite Communications System. In short, the extensive network of Western military satellites used for communication, surveillance, early warning, navigation and geodesy represents a co-operative allied effort. The United States plays by far the largest role but even many US national space assets have an important European operational input.

The use of the satellite network and the information it generates are also shared to some extent by the Western allies. It is now standard practice for the member nations of NATO to enter into mutual support arrangements which allow access to another nation's satellite facilities and information. However, there is an inevitable desire for nations to have independent control over assets that have such a high inherent capability to support military operations—a desire reinforced by a current shortage of satellite capacity within NATO and the reluctance of the largest gatherer of satellite intelligence, the USA, to give access to the most sensitive data. For this reason there has been a considerable proliferation of *national* surveillance and communications satellites. It has already reached the point where the United States may change its 'open skies' policy now that allies are deploying systems, many of them commercial, capable of providing multispectoral, multisensor data which gives information of military quality.[10] Nevertheless, America's allies receive much of the intelligence information on the European theatre from US reconnaissance satellites and use some of its communications channels. Britain also gets guidance data for the targeting of the Polaris force. Even so, dependence on US satellites in the early phases of a European conventional war is likely to be low given the abundant sources of communication and reconnaissance facilities within the theatre, many of them under the national control of European governments. However, should a central-front war be prolonged and 'in-theatre' assets destroyed, satellites would play an important role in the prosecution and, even more importantly, the termination of the war.[11]

The greatest shared interest in space assets concerns the degree to which they underpin the effectiveness of NATO's conventional and strategic policies. Their value for theatre and tactical conventional war—the force multiplier effect—has already been noted and the roles include assessments of the size and readiness of opposition forces, the location of air defences and a growing importance for antisubmarine warfare. Systems like NAVSTAR will improve the accuracy of firepower down to platoon level and will become even more important if NATO implements a conventional strategy based on follow-on force attack (FOFA) which requires the location and strike of second- and third-echelon Soviet forces in the Warsaw Pact's rear areas.[12] At the nuclear level satellites play a vital part in early warning, targeting and guidance of strategic and theatre nuclear forces as well as the provision of 'terrain comparison' data for Cruise missile guidance. As Smith has shown elsewhere in this volume, these are central to the extended deterrent upon which NATO security ultimately rests.

These shared interests in space are highly valued by the European members of NATO, which see the current passive and 'peaceful' use of space as an essential underpinning not only of NATO's defence and deterrent posture, but also of a stable East/West relationship. They have therefore been reluctant to see the 'weaponisation' of space and have pressed for the conclusion of a treaty to limit the deployment of ASAT technologics. However, even improvements in passive satellite capability can be threatening and there is currently concern about Soviet satellite developments that could influence the outcome of a European war.

The allies are concerned about the Soviet's ROARSAT capability to locate US shipping, which could be vital in any battle to keep open the Atlantic reinforcement bridge. An improved satellite capability for battle management may also provide Warsaw Pact forces with a mobility and flexibility currently denied them by a very rigidly centralised command structure and, if this were so, the threat posed by Soviet operational manoeuvre groups (OMGs) would be increased. NATO is also exercised by the threat of fast-improving Soviet satellites used in conjunction with dual-capable tactical missiles; these provide the Warsaw Pact with its own deepstrike conventional options as well as a theatre nuclear capability with considerable options for selective use. All this makes the satellite assets of both sides a tempting military target and forms part of the explanation of the superpowers' interest in developing an ASAT capability.

ANTISATELLITE WEAPONS AND EUROPEAN SECURITY

ASAT weapons are not new. The United States deployed its first system in 1962 and a second on Johnson Island from 1963 until 1975. The latter was abandoned because it had serious operational limitations[13] and because a more promising system, the MHV, was already under development. The MHV began life as part of America's early research into anti-ballistic missile defences but was redesignated an ASAT programme following the signature of the ABM Treaty in 1972.[14] It is now the core of the current American ASAT effort and provides a capability that is more flexible, more versatile and more effective than the co-orbital system deployed since 1968 by the USSR. That system has been untested for the last three years and General Lew Allen testified to the Senate Foreign Relations Committee in 1979 that it has a 'very questionable operational capability'. Despite this and a Congressional amendment to the 1984 defence budget restricting further US tests, the Reagan Administration has placed a high priority on developing an ASAT capability. Its position remains that outlined in the President's report to Congress on 2 April 1984 that 'no arrangements or agreements beyond those already governing military activities in outer space have been found to date that are judged to be in the overall interests of the United States and its allies'.[15]

However, the overall interests of the United States are very different from those of the European allies. The Reagan Administration's policy states that a US ASAT programme is necessary to match Soviet capability and also to deter Soviet attacks on American satellites. The logic of these positions has been questioned by European governments. Most authorities assess that the United States has a considerable lead in ASAT capability and Europeans fear that further development will encourage the USSR to close the gap rather than allow the United States to widen it. It is doubtful that ASAT 'deterrence' makes much sense either since that must rest up on a rough parity of dependence on satellite assets; but the degree of superpower dependence varies considerably from one satellite mission to another and from one theatre to another, although overall the US is thought to be the more dependent.[16]

These differences have counted for relatively little because the administration's real objection to an ASAT treaty is that it would almost certainly slow down the development of the SDI programme.[17] The administration prefers the present situation, which allows many kinds of BMD testing currently banned by the ABM Treaty to be conducted under an ASAT test guise. As two staunch supporters of the administration's policy have put it:

if the United States agreed to ASAT deployment and test limits, it would thereby forfeit or postpone achievement of the distant goal of providing effective defences for the continental United States. ASAT development will provide the stepping stones toward BMD . . . One cannot ban ASAT weapons without, *ipso facto*, banning effective multilayered BMD.[18]

Even so these 'stepping stones' might prove to be treacherous because an effective multilayered BMD depends upon both 'birth to death tracking' by satellite and the use of highly vulnerable space-based attack systems, and these are precisely what an effective ASAT capability would threaten. The Office of Technology Assessment has calculated that the satellite and space platform defensive technologies currently under development by the DOD are always likely to be outstripped by new ASAT threats.[19]

These factors, especially the link with BMD, serve to reinforce European interests in limiting the development of ASAT capabilities by both superpowers. In addition Europeans fear that deployment will extend an already unhealthy arms competition into outer space without giving a decisive advantage to the West. Indeed, ASAT deployment will undermine the superpower strategic relationship and Western interests by threatening mechanisms for the control of nuclear weapons. ASAT would not affect either side's ability to launch a nuclear strike or to reply, but it could destroy the capacity for controlled escalation and seriously complicate the problem of terminating a nuclear exchange. In addition to this the promotion of ASAT weapons may undermine the ABM regime and under these circumstances offensive systems would be even more dominant in space and strategic weaponry than at present.

On the conventional battlefield ASAT might also produce problems for the Europeans. There is little danger of ASAT capabilities supporting a surprise attack by either side since their use would give sufficient warning of a ground attack. But they will threaten command, control and communication facilities during war. These are the principal technological devices possessed by both sides that offer a slim chance of keeping a conflict on the central front at the conventional level. With ASAT there would always be the risk that the side losing the conventional battle would use it in an attempt to 'blind' the adversary in order to regroup or redirect its attack. But in so doing it would increase the risks of uncontrolled escalation. Yet, it is precisely this kind of strike that the Reagan Administration has in mind. In its April 1984 statement to Congress it said:

> The US ASAT program is focused explicitly on those Soviet satellites which threaten US and allied terrestrial interests in time of war . . . While the US MV ASAT will be able to attack only a portion of Soviet satellites, in so doing it would be able to seriously degrade the Soviet reconnaissance capability.[20]

ASAT deployment would also confront those European states that operate their own satellites with either the risk of losing them or the considerable cost of undertaking protective measures.[21] Sir Geoffrey Howe spoke for all Europeans in March 1985 when he stressed the urgency of the superpowers' reaching an agreement in the near future to prevent the deployment of ASAT—in his words: 'In the case of anti-satellite systems, the future is now'.[22] It is notable, however, that this was an aside in a speech which focused upon, and remains the most cogent expression of, European disquiet about the SDI.

STRATEGIC DEFENCES AND EUROPEAN SECURITY

European governments share the deep-rooted scepticism of much of the scientific community about the technological feasibility of the SDI, and doubt that it can ever be made cost effective given the inherent vulnerability of space-based systems.[23] Some of these issues are covered in other chapters and, in any event, will not be resolved for several years. It is intended, therefore, to concentrate on the immediate implications of strategic defence programmes for the European members of NATO. These are of three kinds: strategic, political and economic.

Strategic issues

From the European perspective, the strategic effects of the SDI can be divided into three categories: those that affect NATO/Warsaw Pact relationships, those that affect the strategic relationship of the United States and Europe and, finally, those that affect individual European members of the Alliance.

The major concern about interbloc relationships is that the SDI will be destabilising. This is a view held almost universally by European governments despite the Reagan Administration's assertions that SDI will devalue offensive arms and encourage the USSR to negotiate seriously for an ordered transition to a stable relationship based on a mix of offensive and defensive weapons.[24] The main reason for European scepticism is the attitude and likely response of the USSR to the SDI. There is little doubt that the USSR regards the SDI as part of an American drive for strategic superiority. The view of Vladimir Katin, a political analyst of Novosti, is typical of a vast body of Soviet literature. He wrote in December 1985 that the picture of the SDI as a defensive system is an 'optical illusion', designed to hide America's attempt to develop weapons for a first strike and to prevent or weaken the Soviet retaliatory strike. Washington, he claims, is prepared to disregard the security interests of allies, 'even their very existence in pursuit of the chimera of military superiority'. Such excursions in Soviet propaganda would count for little were it not for the fact that many Europeans, the Secretary General of NATO not least among them, hold views that are uncomfortably close to these.[25]

Because it is widely held, even by the majority in the Reagan Administration, that BMD cannot be made perfect, analysts argue that strategic defences would work best against forces already depleted by a nuclear strike. It is this fact that led General Brent Scowcroft to testify before the House Armed Services Committee that a workable Star Wars system might actually be an 'inducement' for a US first strike.[26] It is this same fact that makes it certain that the USSR will attempt to counter the SDI threat to its ICBMs which, with some 70 per cent of the total warhead inventory, are the very essence of Soviet strategic power. Precisely what form that counter will take is difficult to predict but it is almost certain that the USSR will attempt both to 'offset' SDI by increasing and improving offensive forces, and to 'emulate' SDI by developing a similar defensive system.[27] The most likely result of this would be, in the words of the normally prudent North Atlantic Assembly, 'a cranking up of both the *defensive* and *offensive* arms races'. Indeed, given the current hypersensitivity of the Reagan Administration to Soviet arms policies one could not rule out a 'shock-driven' programme for the crash deployment of the SDI.[28]

The major concern about intrabloc relationships is that the SDI will cause a strategic decoupling of America from its European allies. The US Secretary of Defense has sought to quell these fears by the frequent recitation of the President's words: 'Their safety and ours are one; no change in technology can, or will, alter that reality'. Declaratory commitments of this kind have not reassured Europeans for at least a decade (as the INF saga vividly demonstrates) and there is even less reason to believe that they will do so now. And statements to the effect that with physical protection 'the willingness of US presidents to run risks on behalf of distant allies logically should be strengthened'[29] will trouble Europeans on two counts: firstly that a President *may* be willing to accept risks that are too high, and secondly that he *need* not choose to take risks at all. In other words, with physical protection for the US, risk on behalf of allies becomes a matter of presidential choice and this too fails to reassure the Europeans. What they seek is a situation in which a threat to the survival of Europe necessarily calls into question the survival of the United States. The position has been most eloquently argued by Christoph Bertram who says that, 'Europeans are profoundly convinced that their security rests on America's recognition of its *own* vulnerability. For Europeans, American–European solidarity is not just a matter of declared interests, but of shared fate.'[30] Europeans recognise, however, that the SDI is in essence designed to reduce America's vulnerability to Soviet ICBMs, and that geography dictates that it is easier to do that than it is to protect Europe from Soviet theatre nuclear systems. With that one must recognise that the respective fates of Europe and the United States are diverging and that present Alliance arrangements may not be flexible enough to serve them both.

The deployment of strategic defences by the USA and USSR would also strain the logic of the Alliance strategy of flexible response and its related concepts of the ladder of nuclear escalation and 'escalation dominance'. Effective *superpower* missile defences conjure up the prospect of a nuclear war limited to Europe, whereas effective *Alliance* missile defences would leave Europe open to the threat of the Warsaw Pact's superior conventional forces. In the first case the ladder of escalation terminates short of the superpowers and in the second short of a nuclear exchange. However, it is more likely that defences will be imperfect, but even this would make the graduated and selected use of nuclear weapons, on which the strategy of flexible response relies, a much more complicated and uncertain exercise than at present. In particular, the initial use of nuclear weapons by NATO would have to be on a much larger scale than presently planned to ensure that they penetrate Warsaw Pact defences. This means that several rungs of the nuclear ladder of escalation would have to be scaled in one go and there is no certainty that that would be credible. A sustained attempt to counter Western defences by an increase in Warsaw Pact nuclear forces would inevitably call into question NATO's ability to maintain escalation dominance.

The strategic policies of Britain and France would be especially hard hit by the deployment of strategic defences, although it is unlikely that their forces could ever be completely nullified. However, both governments may have to undertake costly programmes to improve the penetration capability of nuclear forces which already press the national defence budgets hard. If the British Trident programme is to be abandoned, it is more likely to be the result of cost considerations rather than the strength of political opposition, but the British government is aware that even a relatively small increase in the cost of Trident may re-open a damaging public nuclear debate. If the SDI were to force Britain into either abandoning Trident or accepting a substantial increase in its cost it would be an especially bitter blow to the special relationship.

It is ironic that the SDI threatens the effectiveness of British and French forces at the same time that it makes them politically more important as a hedge against an American isolationism which strategic defences imply. For the French government this reinforces their view both that the American nuclear guarantee cannot be relied upon absolutely, and that French nuclear forces have a growing role to play in the security of Western Europe.[31] For these reasons both Britain and France are more likely to pay for an improved strategic performance by cutting conventional forces. A proposal to reduce the British Army of the Rhine (BAOR) by 50 per cent has already gained currency in the United Kingdom and, even though it is not a plausible option in the short term, views like these may yet confound Alliance plans to improve conventional capability.[32]

Political issues

The political implications of the SDI for Europe are the most immediate and fall into four broad categories: Alliance procedures for consultation in peace and war, the opportunity cost of the SDI, its effect on public opinion and the 'elite defence consensus', and a myriad of concerns about arms control. President Reagan's Star Wars speech was made without prior consultation with his major allies and it is clear the European security interests were 'tacked on' to the speech at the last moment. These issues still vex European governments.[33] However, faults in the procedures for peacetime consultation pale into insignificance when compared with the situation that would exist in wartime if SDI were part of NATO's defence arrangements. James Fletcher describes the kind of system that might be used for US and Allied defence as one that has a 'semiautonomous battle management' capability that 'in most cases replaces human decision making'. Even a system designed solely for the defence of the United States might require bases in Europe for 'pop-up' weapons that would engage Soviet missiles on the command of a similar computer-based control system. There is little enthusiasm for this kind of system in Europe.[34]

The opportunity cost problem is twofold: firstly that the US may divert funds from its conventional forces into the SDI and so do less in NATO, and secondly that the Europeans will be required to carry even more of the Alliance conventional burden. Neither the Europeans nor General Bernard Rogers feel that the latter is realistic, especially if the Europeans themselves become involved in SDI work. The military budgets of all NATO countries are severely constrained and a heavy investment in SDI is bound to divert attention from conventional defence needs. This would again call into question the US nuclear guarantee, and also render void Alliance plans for conventional options designed to raise the nuclear threshold.[35]

The desire to raise the nuclear threshold is itself an attempt to placate public concern about NATO's dependence on nuclear weapons, but SDI threatens to rekindle that concern for a number of reasons. The first is the President's insistence on describing the present Alliance policy of retaliatory deterrence as immoral. European governments fear that this will destroy the fragile public acceptance of Cruise and Pershing II missiles which was so narrowly and so hard won in the early 1980s. The second is that the SDI may itself provoke public opposition, in part because it has a nuclear element of its own and in part because several European governments, many opposition parties and large sections of the public regard SDI as the major barrier to the conclusion of arms control agreements. It has been especially difficult during recent months for European governments to support the United States publicly when the USSR has been making the pace in setting an arms control agenda that many in the West regard as constructive, but which has been met with the opposition of the Reagan Administration. These problems are dealt with in detail in the next chapter, and are at the heart of the SDI-induced political differences that divide the trans-Atlantic partners of NATO.

Economic issues

These strategic and political differences have been muted and submerged to some extent by European participation in the research and development aspects of the SDI programme. This is in part a 'solution' to a perceived economic problem. Its essence is the Europeans' belief that the $25 billion or so that will be invested in the first five years of the SDI programme is a disguised industrial policy that will fund a massive expansion in America's high technology industries that will give them an equally large competitive advantage in military and civilian markets. Competitive advantage in the civilian marketplace depends to a great extent on the degree of spin-off that can be expected from SDI work, and is the subject of a heated controversy.[36] However, the threat to Europe's commercial competitiveness has been taken seriously enough to persuade the French to launch the EUREKA (European Research Co-ordination Agency) programme and the European Communities to develop plans for a 'technology community'—both designed to parallel SDI investment in technologies that are assessed to have considerable potential for commercial exploitation.[37]

The viability of these programmes is still in doubt and in no small measure as a result of the US invitation of March 1985 for European participation in the SDI research programme. The offer has produced a deep ambivalence among European governments in that their opposition to SDI on strategic and political grounds remains as strong as ever, but is not voiced while there is the prospect of valuable SDI contracts. Only the French, Greek and Danish governments have formally rejected association with the SDI, although the participation of private industries has not been ruled out. The United Kingdom and the Federal Republic of Germany on the other hand have both signed agreements with the United States, and other European states have adopted an intermediate position. Participation has been an important element in the Reagan Administration's battle with Congress for SDI backing, and has been successful in containing European criticism of SDI. But participation raises considerable problems of its own.

In the forefront of European concerns is the very stringent US approach to technology transfer, and the imposition by the US of 'more than a normal measure of control' over SDI contracts to prevent key technologies being passed to the USSR.[38] *Jane's Defence Weekly* noted in February 1986 the American attempt to study and classify the European high technology base, and described this as 'a dramatic departure in US policy on high technology', by which the US is becoming 'pro-active' in attempting to anticipate technologies that may have a critical military value and to prevent or delay them reaching the marketplace. This would amount to a US veto on

Europe's capacity to compete commercially in vital areas like information technology, optics and high energy lasers. The US study has been commissioned by Paul Hopler, chief technology analyst for the DOD, who is responsible for maintaining the Critical Military Technologies List. The proposed study has already caused problems between Britain and America after the Ministry of Defence refused Hoplers' unofficial study team permission to impose security classifications on British companies.[39]

It is understandable that European countries will not allow this kind of intervention into their high technology industries given the extreme difficulty of obtaining secure and substantial participation in the SDI programme. The head of the British SDI Participation Office has talked of the 'almost insurmountable trouble' in securing SDI contracts and some estimates place the total sum on offer to Europe at as little as $30 million over the next five years. SDI participation will not offer the Europeans a technological panacea, and even those contracts that are secured are likely to produce relatively little commercial spin-off.[40] The participation 'solution' is unlikely to hold the Alliance together on SDI for very much longer. But there is another candidate in the offing which may yet be all things to all allies, and that is a European Defence Initiative (EDI) built around an ATBM.

TACTICAL DEFENCES AND EUROPEAN SECURITY

There has been much political interest generated during recent months in the idea of developing an ATBM for Europe. Initially the Reagan Administration argued that the SDI would be as equally effective in defending Europe against Soviet short and intermediate-range missiles as it would be in defending the United States against ICBMs. In 1984 Dr Edward Gerry, a member of the Defence Technologies Study Team, argued that boost-phase systems 'would be equally effective against intermediate or theatre nuclear ballistic missiles, or even against tactical ballistic missiles.' Richard DeLauer, Under-Secretary of Defense for Research and Engineering, supported this analysis and, almost in passing, swept aside the problem of very short flight times by suggesting that this would be compensated by the fact that in Europe 'the impact area is very limited.'[41] These arguments failed on two grounds; the first was that the Europeans did not want the SDI even if it could offer some defence for Europe, and the second was that SDI addressed defence against strategic missiles alone and there was no, nor is there, evidence that such systems will perform well against a tactical missiles threat. If Europe was to be defended another rationale and another system had to be found.

The rationale was not long in coming and that was the need to find a defence for NATO against the *conventional* use of Soviet dual-capable missiles like the SS-21 and SS-23. General Bernard Rogers, NATO's Supreme Commander, and General G. C. Berkhof of the Dutch army have described how these systems could be used to suppress NATO's air forces, cripple its communications network and 'pave the way for high-speed "Blitzkrieg-type" operations by Operational Manoeuvre Groups.'[42] Their solution is a theatre-specific ATBM, the most likely candidates for the role being either the improved Patriot or the improved Hawk missile, both manufactured in the USA. Neither are formally part of the SDI programme even though they possess some defensive capability against ballistic missiles, but the links to SDI in the thinking of the Reagan Administration lie very close to the surface.

The European members of NATO are considering ATBM with Manfred Worner, the Federal Republic's Defence Minister, as its most active apostle.[43] There is a genuine interest in tactical missiles defences as a protection against the scenario described by Berkhof and others, but there are two additional potential benefits. The first is that European support for ATBM may be an acceptable alternative to joining the SDI proper, and the second is that participation in an ATBM programme might benefit the European high technology base more than an equivalent level of participation in SDI. However, these benefits will accrue only if the ATBM programme is an essentially European one, and this has encouraged Worner and others to think terms of developing a European programme and missile to fulfil the role. This, however, has raised Alliance problems of its own.

Political issues

At the March 1986 meeting of NATO's Nuclear Planning Group, General Abrahamson, head of the SDIO, indicated that European 'theatre defence' would be incorporated into SDI. The Reagan Administration wants the ATBM programme to be seen as part of the SDI for political reasons and also wants to involve US weapons systems that can be developed and tested in a way that will drive the SDI programme forward. This has the twin advantages of circumventing of current arms control limitations on ABM testing, and developing technologies which in the longer term could provide some defence for the continental United States. For these reasons funding for an ATBM has been part of the US defence budget since 1982, and the DOD has sketched out a role for the system in its NATO study 'Counter Air 90'. Patriot and Hawk missiles will soon be tested against short-range ballistic missiles, and the SDIO has announced plans for 'systems demonstrations' with ATBM applications for the early 1990s.

These developments confirm the US interests in tactical missile defences clearly stated by the Hoffman Panel. It recommended a NATO ATBM as one of the intermediate options that could be available relatively early, suggesting that its components 'although developed initially in the ATM mode, might thereafter play a role in continental United States (CONUS) defence.' It made the additional points that this solution would address 'the pressing need to protect allied forces, as well as our own, in theatres of operations from either non-nuclear or nuclear attack' and that it, 'should reduce allied anxieties that our increased emphasis on defence might indicate a weakening in our commitment to the defences of Europe. We can pursue such a program within ABM Treaty constraints.'[44]

All this makes clear that the Reagan Administration has made every effort to ensure that ATBM remains closely locked into the SDI and that European alternatives to these programmes will not be welcomed.[45] An American ATBM option clearly has the merits for America of promoting the sales possibilities of at least two of its weapons systems, providing intermediate technology that will drive the SDI programme, buying off European anxieties about decoupling, and allowing the testing and development of component parts of the SDI without infringing the constraints of the ABM Treaty. Little wonder that it is so vigorously pushed by members of the Reagan Administration and Republican members of Congress.[46]

Nevertheless, European governments, especially those of the FRG and the UK, wish to see ATBM develop as a largely European programme that demonstrates support for SDI in principle, but does not become part of it. In a press release issued immediately after the March 1986 NATO Nuclear Planning Group in which the US declared European theatre defence to be part of the SDI, George Younger, the British Defence Minister, said that European NATO leaders envisaged a high technology defence umbrella linked to but *independent* of SDI.[47] Independence would ensure that any future deployment of ATBM programmes would not be seen as an endorsement of the deployment of SDI, and also ensure that the collapse or postponement of SDI would not take ATBM with it. European governments regard this as important because ATBM, unlike SDI, is intended to counter an existing and immediate threat which is not currently the subject of arms control negotiations. European statesmen seem to be clear that if the ATBM option is taken up care must be exercised to detach it from, and to prevent it from giving support to, attempts to deploy the SDI.

Political/economic issues

David Greenwood has spelled out in great detail the economic advantages of an Initiative for the Defence of European Aerospace (IDEA) based on an ATBM programme that would match the SDI, 'if not in scope and scale, at least in broad conception and purpose, including the purpose of providing a powerful (even urgent) impulse to technological advance'.[48] He sees it not so much as an alternative to more ambitious high technology schemes like EUREKA and the EC's 'technology community' programme, but as a more coherent and precisely targeted endeavour likely to mobilise European states more readily than these other options. However, the proposal is rather optimistic about intra-European political co-operation.

Greenwood proposes a task force to elaborate a strategic concept and research agenda for an ATBM programme and an IDEA Co-ordination Office (IDEACO) to manage the resulting multinational research effort. But there is little past or present evidence that the European members of NATO are capable of this kind of sustained collective endeavour. Berkhof is even more ambitious in proposing the creation of a NATO Aerospace Defence Command (NADC) for a combined defence against aircraft and missiles to be headed by a West German officer, and seems to assume that such an arrangement will be both politically acceptable to other European states and strategically credible to the USSR, but offers little supporting evidence.[49]

Strategic issues

An ATBM system is likely to run into many of the military objections raised by European governments against US strategic defences. The first is that in the absence of a negotiated or very limited introduction of ATBM, the Warsaw Pact would almost certainly seek both to counter and to emulate the programme, especially if it were seen as part of SDI deployment. This would stimulate a defensive and offensive arms race at the tactical level. The Soviet attempt to counter need not be restricted to tactical ballistic missiles but could include increases in Cruise missiles, nuclear-capable artillery and high performance aircraft. Therefore, like BMD, the success of an ATBM system depends ultimately upon Warsaw Pact forces remaining static, and this is possible only in the context of an arms control agreement. Certainly the ability of the Europeans to win a defensive/offensive arms race seems to be out of the question. The *Economist* already believes that the estimated $30 million cost of a modest Patriot-based system is too much for European governments to bear and sees no possibility of Europe being able to afford a more extensive system.[50]

Should the Warsaw Pact also emulate NATO and develop its own ATBM capability then the logic of flexible response and extended deterrence would again be called into question. These policies require the US President to have a wide range of options for the selective use of nuclear forces to ensure 'escalation control' and the ability to terminate a nuclear conflict. But even a partially effective Warsaw Pact ATBM system would drastically cut those options and undermine control. Without these options the credibility of the US guarantee to Europe would be called into question and this could make it difficult to gain support for the guarantee in Congress and from the American public. A further difficulty is that the current proposals for ATBM are couched in a division-of-labour context with the USA assuming primary responsibility for strategic defences and the Europeans the responsibility for tactical defences. These programmes, already poorly linked in political terms, have an even less certain connection in strategic logic. If an ATBM system were self contained technologically there is no reason to suppose that an attack that engaged it should imply an attack upon the US or its strategic defences. This means that the SDI/ATBM split may simply revive the decoupling issue in a different form. All these considerations make tactical missile defences for Europe a very difficult prospect, and will require European governments to tread a very fine line between developing a more secure military posture in Europe on the one hand and preventing the incremental deployment of SDI programmes on the other.

CONCLUSION

The militarisation of space has focused upon President Reagan's policy for strategic defence, but its development has come to involve the related areas of ASAT weapons and ATBMs. These new weapons systems will be developed in some form, and together they will change the strategic environment for superpowers and allies alike. It is imperative the European members of NATO develop positions and policies that will support regional defence interests to the fullest possible extent.

There are no regional or national interests in Europe that will be served by the development of ASAT weapons by either superpower, but European governments can do little beyond press for a superpower treaty to limit further ASAT deployment. Should that fail, it is imperative that the US mobile F-15/MHV ASAT system should not be deployed in Europe. This, at least, would be a political message to the USSR that the European members of NATO wish to see a capability for crisis and conflict control survive on both sides of the European theatre.

Much the same conclusion has been reached in Europe about the SDI. The North Atlantic Assembly has said: 'From a purely European standpoint it still has to be demonstrated why a mixed offensive–defensive strategy would be better for Europe than, for example, a strengthened ABM Treaty and deep cuts in strategic offensive weapons'.[51] Here, European governments should at least press the United States to honour the terms of the Camp David agreement, that is to confirm that deployment of any SDI system or component should be a matter of allied and superpower consultation. This would lend European support to the reviving possibilities of superpower arms control agreements that could make much of the SDI politically unnecessary or strategically innocuous. If SDI becomes a barrier to arms control, it is vital for states like the United Kingdom and the Federal Republic of Germany to reconsider the benefits of SDI participation, especially if there are other European programmes that offer an equal or better economic return.

The most complex set of issues for Europe to arise out of the current phase of the militarisation of space relates, paradoxically, to ATBM. There are real military benefits to be had from a European system that confronts the dual-capable threat of SS-21s, 22s and 23s, and other Soviet 'deep strike' weapons. But some measure of theatre balance must be achieved, whether negotiated or not, if a Warsaw Pact counter designed to overwhelm the system is to be avoided. The balance will be easier to establish if ATBM is *not* perceived to be part of an SDI deployment. However, this is precisely what the Reagan Administration requires and is promoting by pushing its own weapons for the ATBM role.

This is a dilemma and may be a catalyst for Europe. The possibility of ATBM being seen as part of SDI will probably force European governments to abandon the dangerous luxury of dealing with Star Wars as a research programme and to grasp the nettle of the incremental deployment of SDI and related technologies. European interests can only be served if ATBM is promoted as part of a regional military balance rather than as part of a programme for the further militarisation of space. That balance also requires that SDI and the ASAT capabilities that it promotes be placed within the context of an arms control regime. The militarisation of space is inevitable, but the SDI is not. NATO members have now to devise policies for the controlled use of space that serve US strategic and European regional security interests. Once again NATO faces a major challenge; once again it is a test of its political maturity rather than its military strength.

NOTES

1. The quotation from *Hamlet* comes from Act II, scene ii, line 263, and the description of the SDI as a 'theatrical dream' from Lord Zuckerman, 'The wonders of Star Wars', *The New York Review* (30 January 1986) p. 38.
2. Cosmo DiMaggio, *The Strategic Defense Initiative: Program Description and Major Issues*, Report No. 86–8 SPR (Congressional Research Service, Washington DC, January 1986) p. 15.
3. *See* Richard DeLauer, 'Antiballistic missile defense—the opportunity and the challenge', *NATO's Sixteen Nations* **29** (November 1984) p. 26, for his description of a notional development schedule for the SDI.
4. This assessment is made in the 'Interim report', North Atlantic Assembly, *Special Committee on Nuclear Strategy and Arms Control*, AD 86, CS/AN (86) (1 May 1986) p. 37.
5. *See* 'The President's Strategic Defense Initiative (excerpts) 3 January 1985', *Survival* (March/April 1985) pp. 79–83.
6. President Reagan's various statements about SDI deployment are covered in 'Interim report' *op. cit.* p. 24, and Casper Weinberger's statement is noted in Sheena Phillips, 'European report', *ADIU Report* **8** (July/August 1986) p. 7.
7. *See* for example General Theodore Ross Milton, 'Space operations center', *NATO's Sixteen Nations* **29** (November 1984) p. 82, who says that the deployment of weapons in space, including SDI systems, is part of an 'irreversible trend'.
8. These examples of UK/US co-operation are noted in a report in the *New Statesman* (31 October 1980) p. 9.

9. Bruce Gumble, 'Air force upgrading defenses at NORAD', *Defense Electronics* (August 1985) p. 103.
10. Douglas A. Rekenthaler, 'World surveillance from space in the 1980s', *NATO's Sixteen Nations* **29** (November 1985) p. 46. The United States is particularly concerned about the French SPOT high resolution multispectoral scanner satellite that became operational in 1985.
11. These issues are covered in Paul Stares, 'US and Soviet military space programs: a comparative assessment', *Daedalus* **114** (Spring 1985) pp. 138–9.
12. *See* Kurt Gottfried and Richard Ned Lebow, 'Anti-satellite weapons: weighing the risks', *Daedalus* **114** (Spring 1985) p. 158.
13. *See* 'Military uses of space' in *Strategic Survey 1985–86* (IISS, London, 1985) pp. 28–36 for a review of this early phase of US ASAT activity.
14. The longevity of the MHV project and the Pentagon's habit of redefining the programme to ensure the circumvention of arms control agreements has been raised frequently in Congress by Representative George E. Brown Jr. *See* for example his comments in the debate on 'ASAT certification special order', *Congressional Record* **131**, No. 108 (Washington DC, Wednesday 4 September 1985).
15. The report is reproduced in 'US will not seek comprehensive ASAT Treaty', *Official Text* (United States Information Service, London, 3 April 1984). *See* p. 2 for this quotation.
16. There is some debate on this issue. Compare, for example, Bellany in this volume with 'Military uses of space' *op. cit.*
17. Ashton B. Carter makes this point forcefully in 'The relationship of ASAT and BMD systems', *Daedalus* **114** (Spring 1985) pp. 184–5.
18. Barry R. Schneider and Colin S. Gray, 'The Soviet military space program', *Signal* **39** (December 1984) p. 73.
19. Office of Technology Assessment, *Anti-Satellite Weapons, Countermeasures, and Arms Control*, OTA-ISC-281 (Washington DC, September 1985) p. 9.
20. 'US will not seek comprehensive ASAT Treaty' *op. cit.* pp. 9–10.
21. This issue was discussed at length in the late 1970s by the Defence and External Affairs sub-committee of the Parliamentary Select Committee on Expenditure. *See* for example 'The future of the United Kingdom's nuclear weapons policy', *Sixth Report from the Expenditure Committee* (HMSO, London, 1979) pp. 76–9.
22. The speech was delivered at the Royal United Services Institute, London, 15 March 1985.
23. Literature on the lack of feasibility of SDI abounds but the best technical survey remains the Office of Technology Assessment, *Ballistic Missile Defense Technologies*, OTA-ISC-254 (Washington DC, September 1985) *passim*, but especially Chs. 7 and 8. Among the best analyses is Hans A. Bethe, Jeffrey Bountwell, Richard L. Garwin, 'BMD technologies and concepts in the 1980s', *Daedalus* **114** (Spring 1985) pp. 53–72.
24. For the official US view *see The President's Strategic Defence Initiative*, (US Government Printing Office, Washington DC, January 1985). Britain's reaction to the SDI and its concern about destabilisation was

given by Sir Geoffrey Howe in his RUSI speech.

25. *See* Vladimir Katin, 'Arguments Against SDI', *Soviet News* (18 December 1985) p. 472, and James Cross, 'NATO legislators overwhelmingly support strategic defence', *NATO Review* (December 1985) pp. 16–21. Cross reports (p. 16) that Lord Carrington argued that the USSR would need to be convinced that any arms reductions would not place the USSR in 'a vulnerable position in which the weapons they had left had been degraded by a successful US deployment of a ballistic missile defence system'.
26. Reported in 'Star Wars developments', *Facts on File* (3 May 1985) p. 317.
27. Stephen M. Meyer, 'Soviet views on SDI', *Survival* (November/December 1985) pp. 274–92.
28. 'Interim report' *op. cit.* p. 36.
29. For one example of Weinberger's pledge on the unity of US and European security *see* 'Weinberger says SDI would benefit US allies', *Official Text* (USIS, 20 December 1984) p. 1. The assessment of presidential willingness to run risks for allies is Colin S. Gray's, 'A case for strategic defense', *Survival* (March/April 1985) p. 54.
30. Christoph Bertram, 'Strategic defence and the Western Alliance', *Daedalus* **114** (Spring 1985) p. 282.
31. *See* for example North Atlantic Assembly, *General Report on Alliance Political Developments*, (NAA, Brussels, October 1985) p. 30, para. 115, and Paul E. Gallis (*et al.*), *The Strategic Defense Initiative and United States Alliance Policy*, Report No. 85–48 F (Congressional Research Service, Washington DC, February 1985) pp. 31–2.
32. Paul Stares, 'The future credibility of the British independent deterrent', *Defence Analysis* **1** (June 1985) pp. 137–9, and Stephen Kirby, *Britain's Defence Policy and NATO* (British Atlantic Committee, London, 1985) *passim.*
33. For a review of European governments' attitudes *see* the papers presented by John P. Holdren and Ernst-Otto Czempiel to the conference co-hosted by the University of California and the Friedrich-Ebert-Stiftung on 'The Strategic Defense Initiative and Western Europe', San Diego, California, 4–6 May 1986.
34. *See* James C. Fletcher, 'Technologies for strategic defense', *Issues in Science and Technology* **1** (Fall 1984) p. 21, and Lord Zuckerman *op. cit.*
35. General Roger's views were expressed in an interview reported in *Jane's Defence Weekly* (April 1985).
36. *See* David Greenwood, 'The SDI and Europe', *ASIDES*, No. 26 (Autumn 1985) pp. 8–32.
37. For a review of these programmes *see* *Bulletin of the European Communities* **18** Nos. 6, 7/8 (1985) *passim.*
38. North Atlantic Assembly, *Report of the Sub-Committee on Defence Co-operation* (NAA, Brussels, October 1985) p. 5, para. 18.
39. Paul Walton, ' "Top secret" SDI technologies ban', *Jane's Defence Weekly* (15 February 1986) *passim*, and Sheena Phillips, 'European report', *ADIU Report* **8** (March/April 1986) p. 9.

40. For a note of the problems of securing SDI contracts *see* the comments of Stanley Orman, Director of the British SDIO Participation Office, in *Aviation Week and Space Technology*, (19 May 1986), and for estimates of the total value of SDI contracts to Europe *see* John Pike, 'SDI contracts: will US allies be sold short?', *International Herald Tribune* (16 January 1986).
41. Gerry's comments are recorded in 'US allies urged to participate in antiballistic missile system', *Aviation Week and Space Technology* (13 August 1984) p. 169. *See also* Delauer *op. cit.* p. 24.
42. G. C. Berkhof, 'The American Strategic Defence Initiative and West European security: a Dutch view' (Conference paper, undated) pp. 7–8. General Rogers' views are recorded in *Atlantic News* (15 January 1986) p. 1.
43. Manfred Worner, 'A missile defence for NATO Europe', *Strategic Review* (Winter 1986) pp. 11–16.
44. Quoted in Pete Wilson, 'A missile defence for Europe: we must respond to the challenge', *Strategic Review* **XIV** (Spring 1986) p. 13.
45. *See* Jonathan B. Stein, 'Anti-tactical ballistic missiles: new threat to the ABM Treaty', *Arms Control Today* (October 1985) pp. 10–12.
46. An ATBM for Europe based upon an American system has been promoted by General Bernard Rogers, Casper Weinberger, Richard DeLauer, Senator Pete Wilson and Representative Duncan Hunter among others.
47. *See Derniers Nouvelles*, NATO Press Service (21 March 1986) p. 1.
48. Greenwood *op. cit.* p. 47.
49. *ibid.* p. 51 and Berkhof *op. cit.* pp. 24–5.
50. 'A Patriot for Europe', *The Economist* (12 January 1985) pp. 39–40.
51. North Atlantic Assembly, 'Interim report', *Special Committee on Nuclear Strategy and Arms Control* AC 189 CS/AN (85) 4 (October 1985) p. 15, para. 58.

10 The Strategic Context of Space

Gordon Robson

The militarisation of space, and particularly BMD as envisaged by the Strategic Defence Initiative has moved the finishing post of the superpower arms race. It is 'the new high ground'[1] of future military strategy. Space has been of growing military significance for some thirty years, contributing fundamentally to the means of delivery, and increasingly to accuracy, reliability, guidance and C^3I facilities. The accelerating pace of exotic technology in the last quarter of the twentieth century has stimulated a rapid increase in all weapons systems and the result is a 'baroque arsenal'[2] in both the conventional and nuclear fields.

Behind this lies a classical conviction that security is a function of the symmetry of countervailing power, whatever the practical difficulties, dubious morality and horrendous consequences. Whilst the 'military balance' defies very precise calculation, the contemporary superpower strategic relationship has become dependent upon a deterrent policy that embraces the threat of MAD. This requires a capability which is of sufficient quality (measured by criteria such as accuracy, reliability, availability and penetration) which is survivable (by hardening, dispersing and hiding)[3] against a 'first strike'. This spectre has grown with technological innovation, such as multiple re-entry vehicles and C^3I, and is supported by a classical preference for 'prevailing' over mere sufficiency or parity.[4]

Arms races are complex processes, 'historically and situationally conditioned'[5] by a range of interrelated forces, from hegemonic aspirations, insecurity whether real or perceived, to the pull of profit and bureaucratic interests and the push of scientific and technological innovation. As a consequence, a feasible proposition today becomes a weapons system tomorrow and one that must be, inevitably, introduced. The pressures which turned Project Manhattan into Hiroshima are to be found in the commitment to exotic contemporary systems, from MIRV technology to the speculative space battle stations. The militarisation of space clearly demonstrates two dominant and interlocking features of the contemporary arms race. Firstly, mutual perceptions of insecurity and asymmetry conditioned by an accelerating technological momentum. Secondly, this arms race is more than a competition in capabilities, it is equally a competition in doctrine, that is the will to win. The relationship between the strategic and technological aspects of the arms race and hence a problem implicit in SDI, is to be found in the words of Harold Brown, ex-President Carter's Secretary of Defense, 'Whether strategy or technology comes first is debatable . . . yet, if we examine how our strategy got to be where it is, clearly technology had a big part in getting us there . . . it is easier to produce military hardware than it is to know what policy to follow'.[6] As the fruits of technical research and development are transformed into space weaponry, so the problem is transferred into the political arena: what to deploy, against whom, where. Whilst technological and doctrinal pressures form the context of strategic choice, the ultimate decision is political, and 'A choice remains fully possible; strategy is guided in part by technical circumstances, but finally it rests upon human intention. It is open to change'.[7]

This chapter will proceed to examine briefly the technological and doctrinal context, and the strategic implications, of the US decision to militarise space further.

Technology and the strategic balance Superpower security is usually perceived in terms of the global strategic balance, although the conventional, European and wider 'out of area' dimensions cannot be discounted. This strategic balance has been of concern to the superpowers since the dawn of the nuclear age and its history is marked by a series of reactions to real or perceived imbalances. Initial Soviet fear of US hegemony has progressed through the age of detente and parity in the early 1970s to mutual fears of 'windows of vulnerability' induced by US strategic modernisation and continuing antagonism. To US strategists, Soviet testing of air defence missiles with an 'ABM mode' and the development of phased-array radars '. . . which consists of thousands of little radars connected so that they sweep the sky electronically . . . a major improvement over the rotating radars . . .',[8] are cited as Soviet violations of the ABM Treaty. Additions to Soviet missile forces such as the highly accurate SS-18 and SS-19, along with the vulnerability of the B-52 to the Foxhound interceptor with look-down radar, and of cruise missiles to the SA-10 surface-to-air missile, suggest that only the Poseidon and Trident leg of the 'triad' may be still credible. If true, the implications of this for Western defence are gloomy.[9] One commentator, Robert Jastrow, offers evidence that present Soviet forces can destroy 70–75 per cent of US Minutemen missiles in a surprise attack and, 'The newest Soviet missiles of the fifth generation, currently being tested in the Pacific, may be able to eliminate 90 to 95 per cent of American ICBM outright'.[10] Paul Nitze, President Reagan's chief arms negotiator, has argued that the Soviet Union has developed a 'prompt hard target kill capability' and that 'they have the only defence against nuclear ballistic missiles today, and they have the only operational antisatellite system, and they have a distinct advantage in conventional weapons on the Eurasian land-mass.'[11]

This perception, and similar Soviet images of Western power, has been responsible, despite problems of evaluation and assessment, for a preoccupation with numbers. As Jastrow observes: 'Experts count and recount missile sites, bombers, submarines, warheads and megatonnage'.[12] The result has been a continual re-evaluation and weapons acquisition, a function not only of mutual hostility and fear, but of the technological imperative. In consequence, as Thee argues, modern military technology has come to 'shape the world of ideas, of fears and hopes' and to determine military doctrines.[13] The application of these new technologies to weapons systems has been impressive, and amongst the most dramatic to impact upon the evolution of nuclear doctrine have been the B-47 jet bomber of the early 1950s, the thermonuclear warhead, ICBMs with MIRVing and highly accurate SLBMs such as the D-5 Trident missiles, and C^3I developments which have demonstrably brought space into the military context. The full militarisation of space will demand the development of both existing and new technologies, although radars, missile interceptors and laser beams have been under development for many years. The new technologies include X-ray lasers, optics and systems to handle vast quantities of fast changing data.[14]

The marriage of modern military technology to the political will for strategic security would seem predestined to militarise space fully and comprehensively. As Jasani has indicated: 'navigation, communications, weather, geodetic, reconnaissance and early warning' satellites have been deployed to enhance military objectives. 'Between 1957 and the end of 1981 some 1,917 such satellites had been launched, which constitutes about seventy five per cent of all satellites orbited.'[15]

All the major nuclear states have developed and deploy satellites in these categories. The US Global Positioning System or NAVSTAR aims to provide a navigation aid which is accurate to within a few metres whilst others of note include US weather satellites such as those of the DSCS located in geostationary orbit, and the MILSTAR system intended for the late 1980s. The centrality of space to US defence is both impressive and quantifiable: 'In the USA, some 70 per cent of all the military long-distance communications traffic is routed by satellites; by 1990, the figure will be 90 per cent'.[16] However, these figures may misrepresent the extent of space militarisation; it is not so much the numbers of satellites that matter but rather their military exclusivity, quality and lifespan.

Against this background of technological innovation, a rough parity characterises the balance of strategic forces in the late 1980s. The superpowers have developed a similar 'triad' although the mix, structure, effectiveness and levels of modernisation differ. The United States was to lead with MIRV technology although it seems now to be qualifying the trend with the development of a small, mobile, single-headed missile (the Midgetman). This obsolescence and continual innovation was noted by Herman Kahn in the 1960s who estimated that a contemporary weapons-system had about five years before technological innovation made it obsolete.[17] This is an extreme view as modernisation, such as the British Cheveline project to update Polaris, prevents total redundancy, and clearly systems such as Minuteman and Trident have been designed for much longer periods. The scale and complexity of BMD indicate a system designed for the next century and one that will clearly absorb and respond to technological developments. Within the last decade all the major nuclear states have been committed to modernisation and for the superpowers this has meant new missiles and manned bombers incorporating increased lethality through multiple warheads, accuracy, invulnerability and 'stealth' technology. Throughout this modernisation programme both superpowers have become more committed to a defensive strategy. This is dramatically represented by Reagan's initial vision of a system to render nuclear weapons 'impotent and obsolete',[18] and whilst 'the objective of a population defence has not been disavowed, the basic objective has now reverted back to the protection of US military capabilities against a Soviet first strike'.[19] In the face of this technological and strategic uncertainty, the Soviet Union has consistently promoted a defensive philosophy. This incorporates civil defence, the 'Galosh' ABM system around Moscow, 6,300 radars, 9,600 missile launchers and 1,200 interception aircraft (compared with 100 US radars and 300 interceptors).[20] This Soviet commitment to defence, and an early interest in space, undoubtedly increases US suspicions and indicates to Payne 'that the Soviet Union is preparing the basis for the rapid deployment of a nationwide BMD system'.[21]

NUCLEAR DOCTRINES AND THE WILL TO WIN

At the very least the objective of superpower nuclear strategy is to survive; at the most it is to win, to prevail either diplomatically or in the nuclear battlefield. As a consequence both superpowers have been engaged in an intense competition which has progressed from a temporary position of US nuclear hegemony in the late 1940s, through stability and parity, towards a first-strike ethos.[22]

The growth of Soviet nuclear capability led Robert McNamara to acknowledge the arrival of mutual vulnerability with stability, parity and nuclear sufficiency that could render unacceptable harm to the opposition the key to security. To McNamara this implied MAD, and that neither superpower could 'attain a first-strike capability in the foreseeable future'.[23] For many within the US the idea of nuclear parity was alien to the conventional wisdom of security through strength. For others it was merely illusionary and transient. Hence, for Theodore Draper, 'Parity cannot be the answer. No-one knows what it means in terms of different nuclear weapons and systems'.[24] However, the most fundamental objection to MAD, and one which signalled the drift towards first strike, was the presumption that it would imply leaving one's country deliberately open to attack, as a watertight nuclear defence would be 'destabilising'. A criticism noted by Henry Kissinger was, 'It cannot often have occurred in history that it was considered an advantageous military doctrine to make your own country deliberately vulnerable'.[25] Similarly President Carter voices the spirit of superiority rather than more equality with: 'I'm not sure that our country has ever been willing, or will be willing, to accept genuine equality, what we define as equality almost inevitably is going to be superiority'.[26]

In 1980, Presidential Directive 59 identified United States strategic policy as being 'countervailing', which Secretary of Defense Weinberger confirmed in 1982 to be the continuing basis for US strategic nuclear policy. This strategy, the ability to retaliate against the Soviet Union at any level, would appear to add little clarity to the strategic debate, representing instead a somewhat fudged view of the sufficiency versus superiority controversy. It signals in fact an increase in US flexibility and capability by placing emphasis upon 'targeting Soviet military and political–governmental command and control centers and also on the destruction of hard military targets'.[27] In essence therefore it is a policy designed to enhance deterrence by threatening the survival of the Soviet system or at least by the denial of victory. This is further emphasised within US strategy by National Security Decision Directive 13 (October 1981), Reagan's strategic modernisation programme (1981) and ultimately, the SDI (March 1983). 'Countervailing', in a word, encapsulates the history of US strategic doctrine, MAD, flexible response and nuclear war fighting. However, it can be argued that it represents a clear signal in US strategic planning that MAD and 'retaliation only' strategies are negotiable.

A new mood and new technological opportunities enable a return to the days of US superiority. The strategic interests of the US are to be maintained by the 'triad' of offensive weapons, although these need not mirror exactly those of the Soviet Union. They must not however become inferior, in either reality or perception, a point noted by Secretary Brown, who argued that 'the political advantages of being seen as the superior strategic power are more real and more usable than the military advantages'. However, Brown was to maintain that 'the era of US superiority is long long past'[28] and two years later the Reagan Administration was to argue that 'the Soviets have attained a margin of nuclear superiority in most important categories'.[29] The renewed concern with nuclear balance has justified strategic modernisation which beginning under Carter has been subsequently accelerated by Reagan, confirming the prospects for strategic instability. Nuclear war fighting, first strike and active missile defence become related and preferred options.

In 1981 the new Reagan Administration began a review of US space options which has clearly reversed the hesitant and incremental pace of previous strategies. Technological and doctrinal considerations obviously underline this response to the perceived Soviet superiority in space weapons. At the technological level, the US was concerned that the Soviet Union was ahead in ASAT development and particularly, despite considerable scepticism, was capable of a directed-energy antisatellite system by the early 1990s. As a consequence the National Security Council was requested to review national space policy in August 1981, and in the same month a Soviet draft space weapons treaty, submitted to UN General Assembly, was rejected. Testing of a new American ASAT began in 1981 and by September 1982 the Air Force Space Command had been established. Doctrinally, these years offered recognition that space provided two strategic opportunities: first the arena from which military force could be projected in order to 'prevail', and secondly, active defence against nuclear attack. Of the former, Paul Stares, writing of the strategic modernisation program, announced by Casper Weinberger on 5 October 1981, argues that the need 'for effective and survivable early warning, communication and attack assessment systems was considered essential if the administration's declared policy of being able to fight and "prevail" in a nuclear war was to be credible'.[30] This option is illustrated further by Stares, who quotes General Robert T. Marsh, Commander of Air Force Systems Command: 'we should move into war-fighting capabilities—that is ground-to-space war-fighting capabilities, space-to-space, space-to-ground . . .'.[31] Also Reagan's speech on 23 March 1983 invited research into BMD and thus in apparent contradiction to the ethos of 'prevailing', stressed the application of technology to the problem of security and defence from nuclear weapons.

Equally, an analysis of Soviet doctrine reflects the ambiguous and twin elements of the deterrent rationale. Both 'sufficiency and superiority' are contained within Soviet strategy and support for each position is to be found in the 'two overlapping but distinguishable spheres: the "socio-political" and the "military–technical".'[32] The former view is usually associated with the political leadership and stresses parity and the mutual vulnerability of the nuclear relationship. Alternatively, strategic principles discussed at the military and operational levels indicate a logic of superiority so that the Soviet Union must be prepared to seize the offensive and to prevail. These need not be contradictory positions, as Garthoff argues, 'Soviet military power, and the constant enhancement of its capability and readiness, is thus justified primarily for deterrence, as well as to wage a war if one should come despite Soviet efforts to prevent it. The view is consistently held by the Soviet military and political leaders'.[33]

To some Western observers used to open debate by civilian strategists, Soviet nuclear strategy remains an enigma. Despite the frequent 'olive branches', the spirit and context of Soviet doctrine often projects an aggressive posture—the dictates of Marxist–Leninist thought rejecting anything but the inevitability of Socialist victory, coupled as it is with the political pre-eminence of the professional military. This marriage of ideology and military culture, against the backcloth of US arms competition, clearly supports the direction of, and gives credence to, Soviet weapons-related technology. Thus it could be expected that the Soviet Union would wish to match Western nuclear forces and if necessary to take pre-emptive action.

Soviet military capability has steadily increased since the Cuba crisis of the 1962 and has been re-emphasised by the US modernisation programme. The Soviet Union is now committed to long-range Cruise missiles, a new intercontinental bomber, code-named Blackjack, new ICBMs such as the SS-X-25, as well as continued deployment of the Typhoon-class submarines. These developments when linked to capabilities in chemical weapons and BMD research provides evidence to the West of the offensive drift in Soviet policy. As a consequence Soviet counterforce has grown since the late 1960s, and 'deterrence via a modern warfighting posture is a significant element in Soviet military doctrine and strategy'.[34] It still remains doubtful whether the Soviet Union has achieved a 'window of opportunity'; the Soviet Union is substantially ahead in strategic delivery systems (2,500 Soviet to 1,950 US) and explosive power (5,800 megatons to 3,800 for the US), but the US remains ahead in strategic warheads (about 11,140 to 8,700 Soviet).[35] These complex and crude figures apart, Aldridge argues that there is no 'evidence that it has the missile lethality, the anti-submarine warfare potential, the ballistic missile defence network, or the space warfare technology to attain a disarming first strike before the end of this century, if then'.[36]

This review of technology and strategic doctrine highlights the pressures supporting the militarisation of space. However, not until the accelerated interest by the Reagan administration, and despite ample opportunities earlier, did the process take-off significantly. In the 1950s and 1960s both superpowers were to have their perceptions of security challenged by space linked developments. For the US, the first Sputnik, the 'missile gap' and Soviet advances in ABM were provocative. Equally, the Soviet Union became concerned at the growing strategic value of US surveillance and communication satellites, and at the development of ballistic defences such as the Nike-Zeus programme of the late 1950s, and by 1969 the Safeguard programme, 'designed to protect American missile silos'.[37]

These developments were to provide a dress rehearsal for the current SDI controversy. When the militarisation of space began in the early 1960s, Hedley Bull could identify the advantages as encouraging stability and reducing 'the degree of inadvertant destruction' by acting as a 'lightening conductor' which draws fire away from the earth'.[38] On the other hand, he noted that the militarisation of space would accelerate the arms race, increase the risk of conflict and inhibit arms control. These now familiar arguments were clearly present in the late 1960s debate over the introduction of a US ABM system. The proponents of that deployment argued that the Soviet Union had such a system, the technology existed and must therefore be exploited and ultimately it could only but save American lives. In contrast the opponents argued that the deployment of a large ABM network would accelerate the arms race, restrict arms control, be less efficient than a 'damage-limiting' strategy of civil defence and ultimately was not technologically feasible.[39] Plus ca change . . .

Throughout these years a number of factors combined to inhibit the militarisation of space, for example shared images of arms control, crisis stability and detente, and technological inadequacies which left satellites vulnerable and ASATs doubtful. Stares argues that the superpowers had tacitly acknowledged the dangers of transferring their adversarial conflict into space. Technical difficulties put question marks over 'relatively unambitious projects such as SAINT, DYNASOR and MOL,' and the demands of alternative weapons programmes, the cost of the Vietnam war and bureaucratic conflict, inhibited full space militarisation. The US Air Force in particular were reluctant and 'wanted a manned space mission first, and an ASAT second'.[40] At this time space had all the makings of a 'sanctuary', at least from weapons, a view supported arguably by the conclusion of the 1967 Outer Space Treaty and 1972 ABM Treaty.

By the late 1970s both superpowers were experiencing pressures which encouraged the move into space. The failure of arms control, such as SALT II, the 'new Cold War' and especially the crisis over Afghanistan, and the growing sophistication of new technologies, had raised the utility of space in both the strategic and conventional fields. The emergent technology of precision guided 'Smart' munitions, and associated military doctrines such as 'Follow-on-Forces Attack', demonstrated the tactical and conventional applications. The battlefield and tactical value of satellites was clearly discernible as they 'offered the chance of improving the lethality of weapons systems and the effectiveness of military forces generally. The net effect was twofold: the dependence on space systems increased, as did the threat they posed to terrestrial forces'.[41] These pressures enabled a return of the 'space warriors', the military chiefs and scientists of both sides, who had originally questioned the wisdom of the 1972 ABM Treaty. In consequence it may be argued that space is moving from 'militarisation', where assets such as C^3I are essentially passive to 'weaponisation' by the new technologies of BMD.

THE STRATEGIC IMPLICATIONS OF BMD

The renewed BMD controversy must be located and analysed within the current structure of superpower relations. The present adversarial relationship is likely to remain unchanged for the foreseeable future; nuclear deterrence will continue to govern the strategic balance, even though the particularities of that balance may change, but the arms control regime offers some minimum ground rules for consensus. With these assumptions noted, the major consequences of a militarised space are threefold. Primarily it challenges the classical ethos of nuclear deterrence—that of mutual vulnerability. In strategic terms, the prospect of BMD diminishes stability, accelerates the nuclear arms race and threatens the momentum of the arms control regime. Secondly, intrabloc consensus would be disturbed by either superpower or bloc leader acquiring a unilateral defensive shield, a problem which has been discussed more within the West where the principle of extended deterrence is crucial to the Alliance. And ultimately, the technological case, its feasibility, survivability and cost effectiveness, remains highly speculative and contentious. The 'space warriors' argue their case in terms of the logic and morality of nuclear prudence, its defence orientation, parity and inevitable technological momentum, the critics point out that to be 'leak-proof', 99.99 per cent of incoming missiles would have to be destroyed, and even if this were possible the strategic consequences could be disastrous.

BMD AND MUTUAL VULNERABILITY

The stability[42] of the nuclear balance is frangible and as with any relationship it is dependent upon perception as much as reality. A condition of strategic instability exists 'if the Governments facing an international crisis feel that it is extremely urgent to decide on the use of nuclear weapons'.[43] This may occur following uncertainties generated by a technical breakout in either offensive or defensive systems, or the prospect of invulnerability from attack. Equally, the vulnerability of SDI assets to Soviet counter-measures could well increase the risk of pre-emption by both superpowers, for the US to safeguard them, and the Soviet Union to prevent this from happening. Despite these greater risks and even if 'peace' can be maintained, BMD could worsen US–Soviet relations, especially during the 'transition to deployment',[44] and therefore escalate further any central adversarial crisis, whilst threatening the entire crisis management regime. To the proponents of BMD, even if imperfect, it should enhance stability by raising the uncertainty and cost of crossing the nuclear threshold. Should BMD develop a tactical missile mode, the whole process could then serve to shift superpower conflict from Europe. This traditional Cold War frontier, for which 'rules of the game' have developed, is at least 'protected' by the nuclear umbrellas of the superpowers. The unpredictable but strategically sensitive Middle East and Africa, and the new frontier of space, become more evidently the arenas for the next generation of superpower conflict. In one respect, however, the present intensified research phase, particularly that of SDI, may be positive—it could prevent an unstable situation arising from a technological surprise or breakout in weapons systems. Also, the US has the added advantage of using the prospect of eventual deployment as a 'bargaining chip' for either arms control or 'good behaviour'. Should the Soviet Union respond with an increase in offensive forces or a carbon copy BMD, the additional economic burden could be calculated as a strategic advantage to the US unless it also compounded pre-emptive pressures.

To initiate any action that could precipitate Armageddon will be the ultimate and most awesome human decision. The reaction times required for BMD, especially during the three to four minute boost phase, will require sophisticated C^3I fail-safes, plus an almost total and unprecedented dependence upon unproven technology and computer systems. At the very least this would seem to imply a 'hair-trigger' launch on warning strategy, if not a full predelegation to machine decision. If, for example, the efficiency of the existing US C^3I system is questionable,[45] the overload induced by SDI will demand a new range of safety procedures to counteract an accidental activation of the BMD mechanism. These include, as at present, lower levels of readiness for periods of relative calm, the exclusion of civilian launch and ICBM test sites from boost-phase cover, and notification of all test launches. The problem of dependence upon machine decision is not new to space defence. Emerging technology and 'Smart' weapons systems indicate the path already being taken within conventional armaments. The US Aegis track-and-fire system, for example, which is currently deployed on four warships, 'will itself determine whether potential surface and air targets have a "hostile intent" and will itself select the "best response", including that of opening fire first'.[46] However, the credibility of this system is open to serious doubts, as Lord Zuckerman observes—Aegis, which 'was designed to track two hundred incoming Cruise missiles, and to engage sixteen of them at the same time, has not yet been shown to be able to manage two or three'.[47]

Whether the current move into space is following the dictates of technology or the lure of parity may not be clear. However, one inevitable strategic consequence would be a fuelling of the nuclear arms race. Therefore 'far from removing strategic missiles, ballistic missile defence would force both sides to find more inventive ways of attacking each other'.[48] To the opponents of space militarisation, the inevitability and momentum of the baroque arsenal will continue as each superpower strives for parity and effective BMD. It is arguable that SDI may generate a totally new kind of arms race, one of 'leaders refusing to accept the limits of military power, of trying therefore to bring military solutions to bear on what are global political and social crisis'.[49] These quantitative and qualitative military pressures will not be restricted to the US and the Soviet Union. The other nuclear states, especially Britain, France and China, will be forced into an increase in offensive arsenals or even space weaponry so as to ensure some penetration, or to accept the spectre of involuntary unilateralism. That the Chinese 'would launch their first astronaut in the next few years and they expect to fly a small re-usable space shuttle by the mid 1990s'[50] may be an augury to the direction of the arms race.

Space technology to date contains the seeds of the 'more inventive ways of attacking', the ASAT weapon. Other systems that are presently under development and that may penetrate a BMD shield include air-breathing 'Stealth' delivery systems, bombers and Cruise missiles, and possibly tactical missiles which may serve to promote an anti-tactical missile defence covering Europe (and for the US the Caribbean basin also?). Should a nuclear exchange occur, BMD could well increase the toll of human casualties and destruction. Whether during the transition to, or after deployment of, a BMD shield, any attacker is likely to attempt to 'overwhelm' the defence by increasing the number and penetrability of existing offensive systems. As SDI, for example, cannot be foolproof, 'the number of Soviet warheads detonating over American territory in the course of an attack could therefore be several times what it would be in the absence of a BMD'.[51]

In the developing folklore of strategic doctrine, the new weapons of a BMD system could be equally deployed offensively. Although 'there is considerable disagreement whether the system would be flexible and likely weapons against targets on the ground or in the lower atmosphere',[52] their antisatellite potential against crucial C^3I assets is not challenged. However, to the opponents of 'Star Wars', space-to-earth weapons (STEW) could be used for first-strike precision attacks upon land targets, both 'hard' ICBM silos and 'soft' C^3I facilities. Kinetic energy and Excimer laser weapons would qualify for this role, assuming that technical problems such as penetrating the atmosphere and concentrating the beams could be overcome. This offensive capability is mentioned in a somewhat contradictory statement by an adviser to the Office of Science and Technology, Colonel C. Harvey: 'The SDI weapons are precisive weapons that destroy only precise targets. They would not make good first-strike weapons because they could not take out the other sides cities, factories and people fast enough. These are weapons which kill weapons'.[53] However, a first-strike would be against weapons and their support structure, and not against cities and people. This would be retaliation. Equally, if space weapons were not used in a first-strike role they would logically complement such an attack by land, sea or air, being 'useful against the second strike of an enemy, many or most of whose weapons had already been destroyed'.[54]

In the original conception of SDI, the purpose was to replace nuclear deterrence and not to enhance it. As the feasibility of a perfect missile umbrella became suspect, a point to be discussed later, so the immediate aims of SDI moved to options which could improve the survivability of existing offensive systems. This, coupled with the US modernisation programme, raises the prospect of invulnerability, real or perceived—in the words of President Reagan: 'if defensive systems were paired with offensive systems they could be regarded as fostering an aggressive policy, and nobody really wants that', and ex-President Nixon: 'such systems would be destablising if they provided a shield so that you could use the sword'.[55] As the US 'sword' already hovers over some 40,000 or more Soviet targets, the 'shield' could offer a formidable addition to the armoury.

Ultimately BMD is likely to compound problems of force comparison and to produce contradictory arms control postures. It was a factor which both encouraged and threatened the Geneva summit of November 1985, which had been encouraged by the failure of INF, the continuing arms race in offensive weapons, US domestic politics and economic pressures, the European peace movement and space weapons. This latter point is noted by Kanter: 'SDI may have had the paradoxical effect both of drawing the Soviet Union back into arms control negotiations and of adding yet another obstacle to reaching agreement'[56]—a view supported by Gromyko who has argued, 'if there were no progress in questions of space, then it would be superfluous to speak about the possibility of reducing strategic arms'.[57] Although President Reagan has denied that the SDI is a 'bargaining chip', it may nevertheless be argued that the prospect of taking the arms race into space has encouraged some arms control co-operation. There is the basis for dialogue through an awareness of mutual vulnerability. Whilst research on space weapons does not violate the essential space agreements such as the ABM, Outer Space and Limited Test Ban Treaties, the testing of devices would. In deference to the diplomatic problems this move would create, President Reagan has recently proposed a seven and a half year moratorium on testing, and the Soviet Union fifteen years. Deployment would serve to dismantle one of the few constructive and co-operative regimes, although arguably the Soviet phased-array radar at Kransnoyarsk and modernisation of the US complexes at Flyingdales in the UK and Thule in Greenland are already violations of the ABM Treaty. The militarisation of space, especially ASATs and SDI generally, remains a fundamental hurdle to the arms control process in the late 1980s.

BMD AND THE ALLIANCE SYSTEM

A strong alliance system is essential to the superpowers' perception of security and the regional balance of power. Further, the principles of extended deterrence and US–Soviet nuclear hegemony, complicated by the British and French independent deterrents, are symbols of leadership and legitimacy, and are instruments to control and manage intrabloc relations. Against this background, the problems of alliance solidarity, particularly those of the West, are compounded by the militarisation of space. Hence Lebow has argued, 'An American commitment to ABM followed by abrogation, *de facto* or *de jure*, of the existing ABM Treaty would magnify enormously the already existing divisions among Europeans and might make NATO's survival a real political question'.[58] To many within Europe, within both the corridors of power and the peace movement, US policy if not malicious, is at least unwise. To the opponents of SDI, the US failed to consult and is pursuing policies which are destabilising and enhance the risk of war, and which could divert resources into an exotic branch of the defence budget and possibly decouple the US from Europe. In essence, a fully protected US heartland is seen as leaving Europe as an unprotected and soft target. To this may be added a particular concern felt by Britain and France, that the credibility of the independent deterrents are weakened or at worst impotent. Finally, the words of C. S. Gray illustrate an unusual possibility in alliance relations if North America were to be threatened by the nuclear actions of its allies, the US may have to utilise 'globally effective (i.e. space-based, rapidly space-deployable, or space platform using) strategic defense architecture for the protection of the planet, regardless of the flag of the attacker or the destination of the attack. The implications of this logic for coalition politics scarcely need elaboration'.[59] The problem is not exclusively military, politically US consensus on strategic defence must hold well into the next century if research is to be funded and deployment permitted. The negativism of Europe could then become either ammunition for opponents, or, further proof to proponents that US security must not be dependent on others.

THE FEASIBILITY OF BMD

The ultimate speculation surrounding BMD is its feasibility: will it work and at what cost? Although it would be naive to assume that no scientific progress, or eventually, deployment, would be possible, the US 'Shuttle' and Soviet Chernobyl disasters demonstrate all too poignantly the fragility of high technology.

The ability to defend against large missile attacks will demand a multilayered architecture that could, 'detect, identify, discriminate, intercept and destroy ballistic missiles'.[60] The proponents of SDI argue that new technologies offer solutions to these problems making BMD a feasible proposition. It is assumed that technological developments over the next two decades will provide complete systems that can identify, track, distinguish (between real weapons, decoys and 'space junk') and ultimately eliminate the threat. To fulfil this objective, ideally at the boost phase of a missile attack, an arsenal of new weapons are under consideration. These weapons, whether they be ground or space based, or 'pop-up', include directed energy weapons such as 'hypervelocity guns', lasers and particle beams. It is argued that these weapons present surmountable challenges, if futuristic visions of the next century, requiring for example lasers capable of being aimed with the accuracy of one metre over several thousand kilometres, and computers capable of performing one billion (US) operations a second.[61] Expressed more graphically, this accuracy requires the ability to strike a stationary target as if one had the 'ability to hit just one particular brick in the Kremlin Wall, using a laser fired from a point in space 200 kilometres vertically above St Paul's Cathedral. In reality, the "brick" would be moving at up to 7 km/sec.'[62]

The feasibility of this vision is subject to four challenges: that it is technologically impossible, would depend upon highly vulnerable systems, could be easily evaded and would not be cost effective. To the critics[63] of BMD, the technology of sensing, interpreting and responding to mass missile attack is not within our grasp. The exotic systems of lasers and particle beams, and the computer controlled 'battle management', will remain untested and unproved for the foreseeable future. The components of BMD, especially satellites,[64] and battle stations, and earth based C³I will be vulnerable to a decapitating counter-attack. Space mines, X-ray lasers and the Soviet Galosh BMD system, which burns out in thirty seconds, could be deployed against SDI space assets.[65] Even prior to SDI, Ball had argued, 'it would require only about 50–100 warheads to destroy the fixed facilities of the national command system or to effectively impair the communication links between the National Command Authorities and the strategic forces'.[66] Should a BMD be deployed, it would have to cope with ICBMs travelling at 7–8 km/s or SLBMs capable of striking targets within five minutes and sophisticated counter-measures employing ASATs, 'Stealth' technology and improved penetration aids and decoys to saturate the defence which would then 'have to face 100,000 or more identical-looking objects during an attack, of which 10,000 will contain real bombs and the rest will be empty decoys', although 'there is not even a viable theoretical *concept* at this time for an effective mid-course defense against such a projected massive threat of thousands of warheads and perhaps even up to a million decoys'.[67] Other countermeasures include low trajectory weapons and fast-burn boosters. 'A 50-second boost phase is feasible, ending when the missile is at an altitude of about 80 kilometres (50 miles). This would make it virtually impossible to attack the missiles during the boost phase'.[68]

The cost of BMD, and SDI in particular, remains an uncertain aspect of the present debate. To the proponents of SDI a cost-effective system is possible, whilst to some opponents a cheaper ground-based system based on the concept of the early 1960s would be preferable. However, studies into cost ratios of the ABM system planned for the 1960s suggest 'for every dollar spent by the Soviets to ensure penetrability of its warheads, the US would have to spend three to ten dollars to offset that improvement'.[69] Cost effectiveness will be measured by comparing the cost to the US of destroying Soviet missiles against the cost to the Soviet Union of increasing their arsenal. The fixed costs of the US BMD, especially the R&D into new weapons, and the C^3I requirements, will make the average cost of each missile intercept considerable. To Gerald Yonas, scientific director of SDI, 'we are working on advanced automation and mass-production technologies to reduce the cost of all the systems we might deploy. We also have a tremendous cost advantage in launching our system into space in peacetime. With advanced high capacity boosters, we can do this much more cheaply than the Soviets can boost warheads into space during an attack'.[70] Also the formula devised by Paul Nitze (although subsequently rejected by the Pentagon)[71] that the SDI must be cost effective at the margin implies that it would cost the US less to maintain and expand the defence than it would the Soviet Union to overwhelm it. However, this is a highly contentious point as 'the technologies required for counter-measures are in much more advanced state than those required for Star Wars.'[72] And to the opponents it is economically absurd, if not morally repugnant, to commit a trillion dollars to a system that is not guaranteed to be foolproof.

On balance, BMD is hardly an offer that we can't refuse. If the imperative of technological determinism is correct we may well have to live with the further militarisation of space. However, it should be noted that technology is limited by the laws of nature and economics, and conditioned by political choice. Equally, the technology which promotes the defensive advance implicit in SDI is the same which is responsible for both offensive and counter-measure capabilities. And one may simply cancel the other whilst leaving a highly destabilising legacy. Whilst the technical feasibility of space defence may be proven one day, the opportunity cost will be high and the impact upon US domestic politics, and the benefit to alliance and superpower political consensus, uncertain.

NOTES

1. *See* T. Karas, *The New High Ground* (New English Library, London, 1984); and the motto of the USAF Space Command (established 1982)—'Guardians of the High Frontier'.
2. *See* M. Kaldor, *The Baroque Arsenal* (Andre Deutsch, London, 1982).
3. For a discussion of these terms *see* G. Prins, *Defended to Death* (Penguin, Harmondsworth, 1983).
4. For a discussion of the Evolution of superpower nuclear relations *see* Union of Concerned Scientists, *Toward a New Security, Lessons of the Forty Years since Trinity* (July 1985) *passim*.
5. M. Thee, *Military Technology, Military Strategy and the Arms Race* (Croom Helm, London, 1986) p. 101.
6. Quoted in C. Campbell, *War Facts Now* (Fontana, London, 1982) p. 164.
7. Union of Concerned Scientists *op. cit.* p. 31.
8. R. Jastrow, *Reason vs. the Scientists: Why the President is Right about Missile Defence* in F. Holroyd (ed.), *Thinking about Nuclear Weapons: Analyses and Prescriptions* (Croom Helm, London, 1985) p. 235.
9. For an assessment and implications of Soviet Power, *see* D. Smith, *Soviet*

Military Power (Armament and Disarmament Information Unit Report, Vol. 4, No. 1, January/February 1982) *passim* and S. Smith, *The Myth of the Soviet Threat* (Royal United Services Institute, Vol. 127, No. 2, June 1982) *passim*.

10. R. Jastrow *op. cit.* p. 238.
11. P. Nitze, quoted in M. Charlton, 'Star Wars' or Peace-in-the Skies' (11) *Encounter* (March 1986) p. 16.
12. R. Jastrow *op. cit.* p. 243.
13. M. Thee *op. cit.* p. 42.
14. For an elaboration *see* S. D. Drell and W. K. Panofsky, 'The case against strategic defense: technical and strategic realities', *Issues in Science Technology* (Fall 1984) p. 56.
15. B. Jasani, *Outer Space—A New Dimension of the Arms Race* (Taylor and Francis, Stockholm, 1982) p. 41.
16. For a general history of space militarisation *see* P. Marsh *The Space Business* (Pelican, Harmondsworth, 1985) Ch. 6; quote p. 114.
17. H. Kahn quoted in M. Wight, *Power Politics* (Pelican, Harmondsworth, 1979) p. 249.
18. President Reagan's Strategic Defence Initiative speech, 23 March 1985.
19. L. Freedman, 'The "Star Wars" debate: the Western Alliance and strategic defence: part 11', *Adelphi Papers* No. 199 (1985) p. 37.
20. Estimates of the Organisation of the Joint Chiefs of Staff quoted in Office of Technology Assessment, *Ballistic Missile Defence Technologies* (Washington DC, September 1985) p. 99.
21. K. B. Payne, *Strategic Defense: 'Star Wars' in Perspective* (Hamilton Press, New York, 1986) p. 60.
22. For a general discussion *see* R. Aldridge, *First Strike* (Pluto Press, London, 1983) *passim*; also M. Charlton, *op. cit. passim*, and D. Frei, *Risks of Unintentional Nuclear War* (United Nations Publication, Geneva, 1982) Ch. 11.
23. R. McNamara, 'Mutural deterrence', *The Times* (24 August 1968) p. 17.
24. T. Draper, 'How not to think about nuclear war', *The New York Review of Books* (15 July 1982) p. 42.
25. H. Kissinger quoted in R. Jastrow *op. cit.* p. 234.
26. President Carter quoted in M. Charlton *op. cit.* p. 23.
27. L. Goure, 'The US "countervailing strategy" in Soviet perception', *Strategic Review* **IX**, No. 4 (Fall 1981) pp. 52–3.
28. H. Brown, DOD FY82 Annual Report, p. 43, quoted in Office of Technology Assessment *op. cit.* p. 79, and p. 43.
29. C. Weinberger, DOD FY84 Annual Report, p. 34, quoted in Office of Technology Assessment *ibid.* p. 80.
30. For a general background to the militarisation of space under Reagan *see* P. Stares, *The Militarisation of Space: US Policy 1945–84* (Cornell University, Ithica, NY, 1985) *passim.* quote p. 217.
31. P. Stares *ibid.* p. 219.
32. Office of Technology Assessment *op. cit.* p. 69.
33. D. Leebaert, *Soviet Military Thinking* (George Allen and Unwin, London, 1981) p. 95.

34. S. M. Meyer, 'Soviet views on SDI', *Survival* **XXVII,** No. 6 (November/December 1985) p. 278.
35. Figures quoted in *Time Magazine* (7 January 1985) p. 38.
36. R. Aldridge *op. cit.* p. 272.
37. A. Chalfont, *Star Wars* (Weidenfeld and Nicolson, London, 1985) p. 98.
38. H. Bull, *The Control of the Arms Race* (Weidenfeld and Nicolson, London, 1961) p. 182.
39. M. H. Halperin, 'The decision to deploy an ABM: bureaucratic and domestic politics in the Johnson Administration', *World Politics*, No. 1 (October 1972) pp. 79–81.
40. P. Stares *op. cit.* p. 241 and p. 242.
41. P. Stares *op. cit.* p. 242.
42. For a comprehensive discussion of the effects of SDI upon stability *see* K. B. Payne *op. cit.*, *passim.*, and P. Stockton, 'Strategic stability between the superpowers', *Adelphi Paper* No. 213 (Winter 1986) *passim.*
43. D. Frei *op. cit.* p. x; and pp. 5–8 for comparative definitions.
44. K. B. Payne, 'Strategic defense and stability', *Orbis* (Summer 1984) *passim.*
45. *See* D. Ford, *The Button: The Nuclear Trigger—Does it Work* (Allen and Unwin, London, 1985) and P. Bracken, *The Command and Control of Nuclear Forces* (Yale, New Haven, 1983) *passim.*
46. W. Kennet, 'Can NATO tolerate "fire-first" rule?' *The Times* (7 August 1986).
47. Lord Zuckerman, 'The wonders of Star Wars' *The New York Review* (30 January 1986) p. 37.
48. W. Burrows, 'Ballistic missile defence—the illusion of security', *Foreign Affairs* No. 62 (Spring 1984) p. 855.
49. J. Manno, *Arming the Heavens* (Dodd, Mead and Co., New York, 1984) p. 171.
50. *See* R. I. Grieves and K. Hendley, 'The new Chinese frontier', *The Times* (30 July 1986).
51. R. N. Lebow, 'Assured strategic stupidity' *Journal of International Affairs* **39**, No. 1 (Summer 1985) p. 70.
52. P. M. Boffey, 'Dark side of Star Wars: system could also attack', *The New York Times* (7 March 1985).
53. P. M. Boffey *ibid.*
54. R. N. Lebow *op. cit.* p. 71.
55. Quoted by Denis Healey, MP, *Hansard* (19 February 1986) clm. 328.
56. A. Kanter, 'Thinking about thc Strategic Defence Initiative', *International Affairs* **61**, No. 3 (Summer 1985) p. 460.
57. Foreign Minister Gromyko quoted by D. Holloway, 'The SDI and the Soviet Union', *Daedalus* **114**, No. 3 (Summer 1985) p. 274.
58. R. N. Lebow *op. cit.* p. 78.
59. C. S. Gray, 'The nuclear winter thesis and U.S. strategic policy', *The Washington Quarterly*, **8**, No. 3, (Summer 1985) p. 91.
60. *See* 'Shooting for the stars', *Time* (14 January 1985) p. 16.
61. J. C. Fletcher, 'The technologies for ballistic missile defense', *Issues in Science and Technology* (Fall 1984) *passim.*

62. R. Buckeley and G. Spinardi, *Space Weapons, Deterrence or Delusion* (Polity Press, Oxford, 1986) p. 107.
63. For a general review *see* E. P. Thompson, *Star Wars* (Penguin, Harmondsworth, 1985) *passim.*, and for some specific technical criticisms, Lord Zuckermann *op. cit. passim.* and S. D. Drell and W. K. H. Panofsky *op. cit. passim.*
64. *See* Union of Concerned Scientists, *Satellite Vulnerability* (Cambridge, Mass., undated) *passim.*
65. E. J. Lerner, 'Star Wars: survivability and stability', *Aerospace America* (September 1985) p. 81.
66. D. Ball, 'Can nuclear war be controlled?', *Adelphi Paper*, No. 169 (1981) p. 35.
67. K. Tsipis, 'Technical and operation considerations of space-based defensive systems' (Paper presented to Symposium, Tuesday 25 February 1986, London). Paper dated November 1985, p. 8.
68. F. Barnaby, *What on Earth is Star Wars?* (Fourth Estate, London, 1986) p. 98.
69. H. A. Bethe, J. Boutwell, R. L. Garwin, 'BMD technologies and concepts in the 1980s', *Daedalus*, Weapons in Space, **1**, No. 2 (Spring 1985) p. 69.
70. E. J. Lerner, 'Star Wars', *Aerospace America* (August 1985) p. 42.
71. See *International Herald Tribune* (3 June 1986).
72. S. Shenfield, 'Soviets may not imitate star wars', *Bulletin of the Atomic Scientists* (June/July 1985) p. 18.

11 Controlling Military Activities in Space

Ian Bellany

The militarisation of outer space and the control of that militarisation have certain special features arising from the nature of the domain. As a medium of military activity outer space lies in an intermediate position between, say, the high seas, on the one hand, with a long tradition of military exploitation, and on the other the Antarctic land mass which has no such tradition and is, of course, the subject of a treaty strictly limiting military activity to a level of zero, except that military personnel or equipment may be employed there for peaceful purposes. Arms, and arms control, in outer space have features which borrow qualities from both extremes, and lie somewhere on the continuum linking the two. Outer space can be compared to the Antarctic when it is perceived as part of the 'global commons'. It is a store of resources, whose outer bounds have yet to be defined, lying outside national territorial boundaries, the object of scientific exploration, and consequently the broad concern of a large number of states. To the extent that outer space is like the high seas it is a potential theatre of war, dominated militarily by the superpowers, whose military activities are limited, where they are limited, chiefly by mutual respect for the other's potential to match whatever extension of military activity of capability the one may be contemplating.

The two arms control treaties in being that limit the military exploitation of outer space, the Outer Space Treaty of 1967 and the ABM Treaty of 1972, owe different amounts to the 'Antarctic' and 'high seas' characterisations. The Outer Space Treaty is the more 'Antarctic' like, lofty and principled in tone, multilateral rather than bilateral, and not a notably immediate check on the military plans of anyone. Its chief clause prohibits the placing of weapons of mass destruction in orbit. The bilateral ABM Treaty is different. It recognises that a degree of military exploitation of outer space is under way—the 'national technical means of verification' consists in large part of space-based hardware—and aims, in Article XII, to protect this from interference. Moreover it explicitly disallows (Article V, para. 1) the development, testing or deployment of ABM systems or components which are space based.

The history of the United States' military exploitation of outer space can also be understood using the 'Antarctic' and the 'high seas' as reference points. The asymmetries or perceived asymmetries in the respective military utility of reconnaissance and, later, communications satellites in earth orbit as between the two superpowers appears to have led the United States to emphasise and promote originally the 'Antarctic' qualities of outer space.[1] If the Soviets themselves had no particular military need for satellite reconnaissance they were unlikely to tolerate the American military exploitation of space for these purposes on a *quid pro quo* basis for very long. The Americans accordingly sought to de-emphasise and underplay their own military interest in space and drew a clear dividing line between their civil and commercial space activities, on the other hand, and the military on the other: the former receiving maximum publicity, the latter minimum publicity.[2]

This policy was successful. Soviet objections in principle to being overflown by reconnaissance satellites ceased to be voiced from about mid 1963. But it is difficult to be certain why the policy succeeded. It is not clear that the division between civil and military space activities in the early post-Sputnik years necessarily meant very much. United States astronauts were exclusively drawn from military personnel. On top of this, the principal military use of space, at the time and now—the gathering and transmission of information—as a 'public good' likely to be of benefit to more than one service branch would not itself have received much stimulus from an individual service branch naturally reluctant to spend its own budgetary resources even partly to the benefit of the other services. In other words the civil programme created the knowledge and experience upon which the military could draw. And civil funding on the scale of the Apollo programme probably pushed the United States up the learning curve of space exploitation faster and more efficiently than if the armed services had had the field left to themselves. In addition the success of the United States' policy was further ensured by the practical problems involved in shooting down satellites.

Satellites in low earth orbit, seen from the surface of the earth, appear and then disappear over the horizon in a matter of minutes. Moreover whilst they may in a twenty-four hour period make regular passes over the same place, successive passes of satellites on the sort of orbits that give maximum surveillance of superpower territory (tilted at a substantial angle to the equator) coming at about ninety-minute intervals will be some hundreds of miles apart. These two factors make, or made, ground-based interception difficult, and by themselves account for the slow progress of the Soviet Union towards a workable antisatellite system. And even this system, based on the launch by fixed-site rocket of an antisatellite satellite that enters an orbit close to that of the target satellite and explodes when the paths of the two cross, is no good for shooting down satellites in higher orbits.

From its position towards the 'Antarctic' end of the continuum United States policy on the militarisation of outer space has gradually moved closer to the other, 'high seas', extremity. This has come about as a result of a decline in the asymmetry of the superpower situations and the creation of a more balanced relationship. This has not arisen because of a balancing of space capabilities; it has arisen because of a balancing of space vulnerabilities.[3]

The Soviet Union discovered a need for reconnaissance, surveillance and communications satellites, for military purposes, rather late. It was not until 1968 that the number of Soviet military satellite launches in a twelve-month period surpassed that of the United States, and even approximate parity of dependence on such satellites was still some years away because of the much shorter average lifetime of Soviet satellites. The growth in Soviet dependence on satellites was a consequence, it must be presumed, of two other factors. Firstly the growth in size and sophistication of the Soviet offensive nuclear forces meant parallel requirements for improved targeting and navigational data of the kind that only satellite-borne sensors could provide. It may also have meant the need for another layer of early-warning technology, based on satellites carrying infrared (heat sensitive) detectors, to supplement ground-based radars, as United States SLBM forces acquired greater range and assumed a greater proportion of United States strategic offensive warheads. The second reason for the Soviet surge of interest in satellites was almost certainly connected with the expansion of the conventional Soviet navy and as an obvious solution to the new communications and surveillance problems associated with naval operations.

The fact of emerging Soviet dependence, of course, immediately opened up a new opportunity for safeguarding United States satellites. Tacitly the restraining factor gradually became an American threat to put itself into a position to disable Soviet satellites if the Soviets sought to put themselves into a position to disable the American.

The United States were in a position to do this quite early. An awareness of their one-sided dependence on satellites provided an immediate defensive motive for investigating how satellites might be shot down. Provided this investigation was not accorded publicity[4] (it had as yet no deterrent role) it did not run a severe risk of creating—through stimulating Soviet interest—the danger it was intended to address. In 1963 the US Army conducted antisatellite tests based upon the Nike-Zeus ABM interceptor of the period from Kwajalein Atoll in the Pacific (the US Army's ABM test site). At the time the Soviets were launching only about ten military satellites a year. By the end of the 1960s the United States had progressed via a short series of test launches (not involving warhead detonation) to a theoretically operational antisatellite system based on Johnston Island, 1,700 miles north-east of Kwajalein, employing converted Thor intermediate-range ballistic missile (IRBM) launchers, with a modified (manoeuvring) upper stage and carrying a nuclear warhead. The operational ceiling of this system appears to have been about 650 km assuming that the test intercept conducted on April 1965[5] against a defunct

Transit 2A US navigation satellite occurred whilst the latter was at its perigee.[6] This system was nominally kept in being until 1975 but never had more than a handful of launchers available for use. Its large nuclear warhead whilst no doubt necessary to overcome difficulties associated with arranging for a close intercept with the target satellite, made it too much a weapon of the last resort to be entirely practicable or credible. There were other, subsidiary, rationales for Johnston Island. It was uncertain how far the Soviet Union retained an interest in offensive nuclear weapons delivered from or *via* earth orbit. The Outer Space Treaty did not outlaw the Soviet fractional orbital bombardment system (FOBS) because it was a nuclear warhead designed to come out of orbit *before* completing its first circle of the earth. The Soviets did not abjure the FOBS until 1972, and there was therefore an overlap between Johnston Island and the United States' ABM programmes of the period. In addition there appears to have been an American wish to add to its ground-based surveillance of space objects, including those with a military purpose, by launching cameras and telescopes into space. It seems likely that the Johnston Island interceptors were also capable of carrying payloads of this kind.

The first Soviet satellite interceptor test series ran from 1968 to 1971, and involved a total of seven tests of a co-orbital intercept system. The actual intercepts took place where the orbits of the attacking and target satellites (usually elliptical and circular, respectively) intersected at or near the point nearest earth, that is at the perigee of the elliptical orbit. Like the US Johnston Island system, although the latter was a direct-ascent interceptor, the Soviet system appeared essentially to be designed to operate against comparatively low orbit satellites (below 1,000 km or so).

The test programme ceased in December 1971, and was not resumed until February 1976. The period between was not only the high point of detente but it was also a period of maximum Soviet–American co-operation in non-military uses of space. Between 1972 and 1975 there were four space ventures involving Soviet and American coparticipation, including the important docking in orbit of the Apollo and Soyuz manned spacecraft in July 1975. The five-year period from 1966 to 1970 had seen no superpower collaborative ventures in space and the period 1976–80 saw only two.[7] The differences between the first Soviet test series and the second do not appear to be fundamental. Available data[8] indicate that the second series, which finished in 1982, involved about the same rate of testing as before (two a year) and a much poorer success rate (about 25 per cent compared with 75 per cent for the earlier series). Another difference, and possibly one that explains the difference in success rates, is much more significant and concerns the orbital height of the test intercepts. The second series looked typically to make intercepts at 1,000 km; the earlier series was content to concentrate around the 500 km mark.

After the standing down of the American Johnston Island system in 1975—although in practice it had been moribund since 1970 when its operational readiness was reduced from twenty-four hours to thirty days—the technical direction of the United States programme altered sharply. After some initial hesitation a new phase of the programme was embarked upon with the choice, in 1978, of a system using the direct-ascent route as before but discarding the fixed, land-based heavy rocket carrying a nuclear warhead in favour of an aircraft-launched light rocket equipped with an infrared guidance system and a conventional warhead—the latter two things constituting the miniature homing vehicle (MHV). This so-called prototype miniature air-launched system (PMAL) was flight tested against a pseudotarget, supposedly successfully, in January 1984, the first American test of a satellite interceptor for fourteen years. The launch aircraft was an F-15. A successful test against a real target (an old US Navy P78-1 satellite) came over a year later, in September 1985, with no lack of publicity.

Whilst actual physical protection of satellites from attack is limited to that which can be done without incurring an intolerable weight penalty, and the emerging Soviet dependence on military satellites opened up, for the United States, an alternative to defence in the shape of deterrence, the vulnerability of a satellite to attack does depend strongly on its orbital altitude. In exactly the opposite to the case with aircraft, where safety from attack is greatest at low altitudes, satellites are safer the higher they are. As a result, the obvious passive defensive measure is to put satellites in as high an orbit as possible. Anything with a perigee in excess of 1,000 to 1,500 km would be out of range of both the Soviet co-orbital system and the new United States air-launched system.

The technical superiority, on paper, of the new United States system over the Soviet co-orbital interceptor is substantial. The air-launched interceptor has a potentially much greater launch rate and is mobile, which both improves its attack flexibility and reduces its vulnerability to attack on the ground.

Other asymmetries also act in the American favour. Arguably the most important and useful of the 'high' orbits, for military or civilian purposes, is the geostationary circular orbit girding the earth 22,300 miles above the equator. Unfortunately this orbit where the satellite appears to hover over a spot on the equator is not directly of use to the Soviets since most of the Soviet Union lies so far north that a geostationary satellite lies below the effective horizon. As a substitute the Soviets have exploited a highly elliptical orbit with a twelve-hour period which permits very slow passes (as seen from the earth) over northern latitudes, and is extremely useful for internal (at least) communications purposes. However, the nature of this orbit, sometimes called Molniya after the Soviet communications satellite, implies that the slow high passes over northern latitudes are matched by rapid but potentially vulnerable low passes over southern latitudes.

The high latitude of most of the Soviet Union creates another geographically derived asymmetry. The natural and economical orbital inclination of a satellite is the latitude of its launch point. Many low orbit United States satellites have an inclination of thirty degrees, the latitude of Cape Canaveral. This inclination does not take the satellite over Soviet territory at all and must add to Soviet difficulties in maintaining a reliable information bank on United States satellite launches and related activities. The natural inclination of Soviet orbits, on the other hand, corresponds to the more northerly Soviet launch sites. For reasons probably connected with a desire to minimise the gyration of a satellite orbit that arises as a consequence of the small equatorial bulge in the shape of the earth, Soviet orbital inclinations are usually about sixty-five degrees, somewhat north of the Soviet Tyuratam launch site in Kazakhstan. The gyration is at a minimum for an angle of 63.4 degrees, close to the favoured Soviet figure. But this also means that all Soviet low orbit satellites at this inclination pass over the United States.

A further minor asymmetry arises as a consequence of a Soviet tendency, particularly marked in the early days of the Soviet space programme, to favour low earth orbits where these were consistent with the mission in hand, whereas the United States has tended to favour higher orbits. Lower orbits mean shorter mission life times, as a result of atmospheric drag on the satellite, which in turn forces the Soviets to maintain a higher satellite launch rate capability than the United States. Whereas around 50 per cent of Soviet military satellites are in low orbit, the United States proportion is only 20 per cent.

These minor asymmetries arising from differences in standard operating procedures when looked at from a military perspective on the whole favour the United States. For instance a military space shuttle could be launched on a thirty-degree orbit with a mission, say, to inspect (telescopically) a particular Soviet low orbit satellite without the Soviets ever becoming aware of it. And low Soviet orbits put Soviet satellites at greater risk from satellite interceptors. On the other hand the higher Soviet satellite launch rate capability does enable them to replace lost satellites promptly—itself militarily useful.[9]

The net effect of the qualitative asymmetries favouring the United States when combined with the quantitative asymmetries in the degree of military dependence on the space medium which still favour the Soviet Union (the United States seems to have about twice as many active military satellites actually in orbit at any one time as the Soviets) can be taken as constituting together a balance of vulnerability.[10] This in turn has led to a balance of deterrence: not a mutual deterrence based on a reciprocal threat to shoot down each other's satellites, but a mutual deterrence based upon a reciprocal threat to acquire a militarily significant capability to do so, were the other side to try and do the same.

The evidence for this—for, in other words, a *de facto* or tacit piece of arms control amounting to a reciprocated hesitation to acquire a militarily significant antisatellite capability—is quite impressive.

Plainly no arms race in satellite interceptors has occurred. Expenditure to date has been small. The Soviet Union has not tested its satellite intercept system since 1982 and in 1983 unilaterally announced a moratorium on further testing. The United States has carried out only one full test of its new air-launched MHV satellite interceptor, which test moreover was sanctioned by Congress only on condition that the Executive certified that it was negotiating with the Soviets to control such weapons. Congressional restrictions after the test were hardened in 1985 to ban any further such full-scale tests by the United States for as long as the Soviet moratorium lasted.[11]

Of course the nature of control has moved from the Antarctica paradigm much closer to that of the high seas.

Vestiges of American interest in control through out and out demilitarisation of space were still to be seen in the United States position[12] in the short-lived antisatellite arms control talks with the Soviets which began in June 1978, and ended without agreement in the collapse of detente, one year later. From the arms controller's point of view the failure of the talks may not have been an entirely bad thing and the tacit arms control arrangement currently in place need not necessarily be thought of as a second best. The reason is that the 1978–9 talks on a formal agreement highlighted a number of difficulties that a formal agreement would of necessity have to tackle.

For one, there is the so-called residual satellite interception capability possessed not only by ABM systems, where they exist, but also by virtually any long-range ballistic missile. Additionally there is the problem posed by dual capable (civilian/military) space launchers, such as the United States shuttle. Furthermore there is the problem of verification: any space-based verification system, problematically, would be next to impossible to distinguish from a space-based satellite interceptor, since to inspect a suspicious object would entail a militarily threatening capacity for close approach, and anyway adversary inspection of a military reconnaissance or communications satellite could itself be judged a hostile act. Arguably a tacit arrangement attains almost everything that could be expected of a formal agreement but more quickly, with far less investment of political effort in arriving at an internally agreed national position, no risk of offending allies or other third parties, and no arousal of excessive popular expectation.

THE STRATEGIC DEFENSE INITIATIVE

President Reagan's strategic defence initiative and the space gloss put on it since the first announcement in March 1983 signified a new phase in the approach of the United States to arms in space.

Arms control is commonly and correctly held to be easier of achievement when it is quantitative; qualitative arms control is harder. The American approach during the Reagan presidency would seem to conform to and ratify this observation. The United States now appears to be most interested in arms control where its effect is to limit numbers of well-established kinds of weapons which the Soviet Union—like the United States—has already demonstrated it can procure at will. It is not so interested in arms control in those less run-of-the-mill areas where the Soviet Union is perceived as technically and qualitatively backward. And by its nature space weaponry falls into the latter category.

The new thrust in American policy simply means that where arms control restraints currently exist on the military exploitation of outer space, these restraints will come under pressure. The restraints are two in number: the 1972 ABM Treaty, and the tacit arrangement limiting competition in antisatellite warfare identified above. The remainder of this chapter assesses the probable holding power of these restraints.

What on this analysis chiefly restrains the United States from stepping over the limits set by the 1972 ABM Treaty is not that the Soviets would respond in kind with a militarily significant ABM system of their own—for that would assume the technological parity whose existence is being denied. Rather the restraining factor is linkage, arising from SALT I and SALT II, between strategic defensive and strategic offensive weapons. The Soviet response would be in offensive weapons and there is little disagreement that a Soviet increase in ICBM warheads of 100 per cent or more could be accomplished on a far shorter time scale than that needed for the United States either to deploy a working modern ABM system, or even to counter-react by increasing its own ICBM or SLBM warhead totals.

It is this linkage that is prominent in the subject matter of the renewed superpower arms control talks which began in Geneva in March 1985 under the three headings: strategic offensive weapons; intermediate-range nuclear weapons; and space weapons.

And what restrains the United States from moving forwards to acquire a fully fledged satellite interception capability is the tacit arms control agreement on antisatellite warfare identified above.

This, standing by itself, would not perhaps be sufficient: the residual excess of American dependence on satellites for military reconnaissance and communication is not now large enough, possibly, to offset Soviet technological weakness, but anything that encouraged the Soviet Union to increase the resources put into satellite interception technology would jeopardise United States' plans to acquire modern ABM defences. Modern ABM defences going beyond the limits set by the ABM Treaty are often depicted as relying on space-based antimissile weaponry, in the form of 'active' satellites that house weapons or act as out-stations for the relay of bolts of directed energy fired from a ground station, with the object of intercepting ballistic missiles in their first, boost-phase minutes of flight. An American investment in such active 'battle' satellites would recreate the one-sided dependence on satellites that existed with respect to informational military satellites in the 1960s and 1970s, and it is hard to see how they could be protected from Soviet satellite interception systems. Whilst battle satellites would presumably be self defended from ballistic missile attack, and would presumably also as far as possible avoid low orbits, neither precaution would be particularly effective if the Soviet satellite interception system was itself of the directed-energy variety fired from ground sites. Again, this is not to ascribe to Soviet military technology in this area of weaponry a degree of sophistication comparable to the American. It is, more plausibly, to suggest first that satellites, with their known orbits and repeated passes, are much easier targets than ballistic missiles and secondly that it is easier for directed-energy weapons to shoot up than to shoot down.[13]

Even steps of a more modest kind toward a modern ABM system could encounter similar obstacles.

One reading of present United States policy is that there is an intention to move ahead in developing SDI technology whilst remaining within the ABM Treaty. The principal avenues for technological development, indeed, are defined by the request of Department of Defense for SDI funding for fiscal year 1987. Most funding is being asked for 'surveillance, acquisition, tracking and kill assessment' and for 'directed-energy weapons'.[14]

The first of these refers chiefly to space-based sensors: 'passive' or 'defensive' sophisticated early-warning and attack assessment satellites that could be described as 'adjuncts' to ABM defence and not disallowed even under a restrictive interpretation of the Treaty.[15]

But as space-based, satellite-borne systems their usefulness would depend on Soviet satellite interception capabilities remaining marginal.

The second priority item on the SDI agenda is directed-energy weapons. Again, the ABM Treaty, even restrictively interpreted, permits the development and testing (Article IV) of fixed-site (ground-based) directed-energy weapons capable of operating in an ABM mode. *Deployment* of such weapons is not allowed. Yet if the United States were to develop such a device it would be virtually indistinguishable from, although more capable than, an antisatellite device and accordingly a stimulus to Soviet developments along similar, even if technologically more modest, lines—especially so, it might be thought, since a treaty-limited test of an ABM has to be carried out from the site designated in 1972 as an ABM test site: in the case of the United States this means the Kwajalein Atoll, where the very first American antisatellite tests were conducted in 1963.

In sum United States SDI activities, even within the ABM Treaty, tend to recreate a one-sided dependence on satellites for military purposes and give the Soviets a strategic incentive to extend their own satellite interceptor capabilities. And directed-energy weapons developed for SDI will give the Soviets technical pointers for the development of the much less technologically demanding and less costly directed-energy weapons that would be good enough for antisatellite purposes. It makes no more sense for a country heavily dependent, militarily, on satellites, to give credence to methods of shooting them down than it would have done for Great Britain in the heyday of the battleship to have 'talked up' the military potential of the submarine.

The element of idealism in the attitude of the United States to the militarisation of space has not altogether disappeared. Echoes of the old Antarcticism can be discovered in the Joint Resolution of Congress (S. J. Res. 236) of October 1984 in support of renewing co-operation in space with the USSR and even in the widespread official dismay at the 'leaking' of the military role behind the Discovery Shuttle launch in January 1985.

But the banner of Antarcticism in space receives its most vigorous flourishing now in the Soviet Union. And the reasons are not very different from those that appealed to the Americans twenty-five years ago: its attraction as a political device to protect a military position that is temporarily somewhat exposed. In this case, for the Kremlin, it is not Soviet reconnaissance and communications satellites that are chiefly seen as at risk (although with a much higher proportion in low orbits than the United States they may see themselves one-sidedly disadvantaged by a rapid build-up in antisatellite weapons): it is the perceived advantage the United States possesses in sensor and directed-energy technology that is the real concern. That this could be translated into a working space-based missile defence network quickly enough to outstrip the counter-development of Soviet directed-energy ground-based methods for shooting down battle satellites seems, on the analysis above, highly unlikely. But whilst a technological gap is seen to exist the Soviets lose nothing by mobilising international opinion against United States plans through characterising the space medium as something to be held inviolate from military competition. And the Soviets are emboldened in taking this line by the slightly curious fact that they have never admitted to the existence of their own satellite interceptor system.

The first public airing of this containment strategy came in a letter from Soviet Foreign Minister Shevardnadze to the UN Secretary General in August 1985. The letter proposed that the agenda of the General Assembly's autumn 1985 session include 'international co-operation in peaceful exploration of outer space in conditions of its non-militarisation', and refers to a possible international (multilateral) conference on the subject. In his speech to the UN on 24 September 1985 Shevardnadze went further:

> The Soviet Union has also submitted to the General Assembly specific proposals concerning the main directions and principles of broad international co-operation in the exploration and use of outer space for peaceful purposes. Outer space is indivisible and all states should be able to take part in its peaceful exploration.
>
> To counter the sinister plans of 'Star Wars' the USSR is putting before the international community a concept of 'Star Peace'.[16]

The Soviet scheme was fleshed out at a press conference later the same day, where a call was made for the renunciation of the militarisation of space, including research to that end, creation of a 'world space organisation' to explore the heavens on behalf of mankind, and an 'international space conference' to plan co-operation among nations with space research capability.[17]

Whilst this is not the only Soviet space arms control card—the Geneva talks offer another route to restraining the United States—its prospects of success are finely balanced. On the one hand the Antarctic approach to the militarisation of space is not entirely dead in the United States, and there is certainly evidence that France—the third space power—at least under its former socialist administration found much to its liking in the Shevardnadze position.[18] Against this there is not only an American wariness and scepticism about a political device that they themselves have used in the past, but also a new degree of insulation between United States' policy and 'international opinion' shown in Washington's emerging preference (especially marked under the Reagan Presidency) for national solutions to foreign policy problems over the collective international variety.

NOTES

1. At first, quite literally: President Eisenhower proposed to the UN General Assembly in September 1960 that the principles of the Antarctic Treaty be applied to outer space and celestial bodies. The object of American diplomacy at this time was less the protection of US reconnaissance satellites and more the hampering of the Soviet ICBM programme—ICBMs because they enter 'space' during their flight did have a brief reputation as 'space weapons'.
2. This is a prominent theme of P. B. Stares' *Space Weapons and US Strategy: Origins and Development* (Croom Helm, London and Sydney, 1985). This book is a gold mine of information and has been extensively quarried in the production of this chapter.
3. The transition of American policy made possible by emerging Soviet military dependence on space is perhaps the chief theme of Stares' book.
4. Publicity was strictly rationed during the period of small Soviet dependence on satellites, but the then US Director of Defence Research and Engineering (Harold Brown) made brief public references to US satellite interception programmes in 1963 and 1964. *See* Robert C. Aldridge, *The Counterforce Syndrome*, (Transnational Institute, Washington, DC and Amsterdam, 1978) p. 54.
5. Stares *op. cit.* p. 124.
6. The orbital parameters of Transit 2A (launched June 1960) were a perigee of 392 miles, an apogee of 655 miles and an orbital inclination (to the equator) of 66.77°. *See* Desmond King—Hele, *Satellites and Scientific Research* (Routledge, London, 1960) p. 170. What is especially noteworthy about this choice of test target is that its orbital inclination comes close to the inclination highly favoured by the Soviet Union for a wide variety of their civil and military satellites (the entire Sputnik series, for instance, had orbital inclinations clustered round the 65° mark.
7. On the other hand it should be noted that in the period 1981–5, in spite of the deterioration in superpower relations, there were four collaborative non-military space ventures, although nothing of the same symbolic magnitude as the Apollo—Soyuz mission.
8. Stares *op. cit.* p. 262.
9. Only for as long as the United States' satellite interceptor had a slow launch rate: this is not likely to be a characteristic of the air-launched system now under development. Moreover even Soviet high orbit satellites are launched *via* a low orbit 'parking' manoeuvre creating a theoretical vulnerability to low orbit satellite interceptors.
10. Whilst figures for respective launching rates, even for military satellites,

Conclusion: Inevitability and Choice in the Militarisation of Space

Stephen Kirby

It is perhaps inevitable that President Reagan's SDI should have dominated recent discussions about the militarisation of space. But in a little over three years the programme has changed so fundamentally that none of the projects currently under development will contribute in any way to the original goals set out in March 1983. The SDI will not protect people nor will it render nuclear weapons obsolete. It will not, as Garden concludes, remove the nuclear threat from superpower relationships. Even the reduced goal of 'enhanced deterrence' is receding as a justification for the programme because there are cheaper and less dangerous ways of reducing the vulnerability of America's ICBMs. Indeed, in the fast-changing political and economic circumstances of America the SDI might already be in terminal decline as a defensible programme. But it has already challenged existing strategic doctrine to such a degree and has accelerated the development of a range of technologies of such potency that the military uses of space are now of first importance in the defence affairs of the superpowers and their allies.

The real importance of the SDI is that it is an accelerator of both US and USSR technology—much of which, as Robson shows, has been around for some considerable time. It is only the development of 'third-generation' nuclear weapons that represent a new SDI-induced technological departure. In the United States these weapons include the X-ray laser, microwave and particle beams and hypervelocity fragmentation device, all driven by nuclear explosion. They have been described as a back-up to the non-nuclear systems of

SDI but are in fact rapidly becoming the major component of the programme.[1] In the Soviet Union similar research programmes could produce 'space-to-earth' technologies and Gorbachev has frequently indicated his willingness to consider weapons deployments in space. Nevertheless, these programmes are still in the early stages of research, and deployment is a distant prospect that may yet be avoided. The SDI has, however, significantly advanced several military technologies that are already deployed and has promoted others that are close to deployment.

These technologies, which include satellite-borne C³I, antisatellite weapons, satellite defence and terminal BMD/ATBM capabilities, require of the superpowers a reconsideration of the strategic philosophy of MAD but do not offer the alternative of an effective defensive astrodome. In Webber's terms the SDI will produce a great deal of hardware but will not produce a system; the computer software problems are simply too demanding, and the counter-measures too numerous and too effective. However, the hardware technologies alone have the potential to underwrite strategic stability or to destroy it. It is inevitable that the militarisation of space will continue, but there is still time for reasoned choices to be made about which space and space-related technologies should be deployed unilaterally, which should be subjected to negotiated limits, and which should be restrained.

These choices must reflect the capacity of each technology to change the existing strategic relationship and the security regime within which the superpowers and their allies interact. Developments like code division multiple access (CDMA) and monolithic microwave integrated circuits (MMIC) have, as van der Veen demonstrates, substantially increased the capacity of satellites both to facilitate communication and early warning, and to provide factual reconnaissance in real or near-real time. Such developments improve strategic and tactical communication, limit the chances of surprise attack, verify arms control agreements, and allow the surveillance of regional conflicts. All these can contribute to the stability of the Soviet–American strategic relationship, especially since technological developments have increased the capacity of military communications satellites to operate in hostile environments.

Many of these developments are required and have been boosted by the SDI programme, but its main requirement is for increased C^3I for battle management for BMD. Smith traces the development of US offensive strategy from counter-force through war fighting to the PD-59 war-winning doctrine and shows how this too requires a massive battle-management capability, a capability which he and Clarke demonstrate to be woefully deficient, highly vulnerable and dangerously destabilising. Even though battle-management capabilities may develop considerably in the near future both the land and space-based components will remain at risk. Lapraik notes the continuing vulnerability of satellites to systems-generated electromagnetic pulse (SGEMP) and to electronic warfare, and ground control and receiving stations for space assets remain few in number and vulnerable to early attack. Even so the proliferation of space-based C^3I could be a strategic 'leveller' and could produce greater stability in superpower relationships were it not for the vulnerability of all space-based facilities to rapidly developing ASAT technologies.

Both the US and Soviet interest in antisatellite techniques are a reflection of the United States' growing dependence on satellite facilities. The demands of SDI in addition to those of war-fighting command and control threaten, according to Bellamy, to re-introduce a one-sided American dependence on satellites that will destroy the current superpower 'balance of vulnerabilities' in space—a balance that has encouraged a tacit form of arms control which, until recently, has inhibited the development and deployment of an effective ASAT system. In the absence of this restraint SDI is dangerous because all BMD developments have an ASAT application. For the United States the SDI will drive ASAT technology, and for the USSR this fact, together with America's growing dependence on satellites, makes an *effective* Soviet ASAT capability a high priority. There are many limitations to the existing Soviet co-orbital antisatellite device and it poses only a small threat to a limited number of US satellites. But much research in the USSR has gone into land-based ASAT technologies using laser and particle-beam principles that, if deployed, would pose a significant threat to all US space-based devices, even satellites in geostationary orbit. Indeed, many of the SDI's laser and particle-beam experiments for BMD will indicate new directions and techniques which the USSR may develop for an antisatellite capability.

Indeed, a major problem and fundamental irony of the SDI is that it will not only produce an effective counter-technology before it produces a working BMD technology, but it will also threaten the C^3I facilities upon which the current US offensive strategy depends. In other words, the development of ASAT technologies denies the possibility of a transition to a 'mutually assured survival' defensive strategic relationship but undermines the existing offensive-based MAD relationship.

McNeill notes the long-term interest of the USSR in developing a military capability in space so that it might alter the 'correlation of forces' in its favour. In this context the SDI has given the USSR an increased incentive to counter and, to some extent, match its technologies. But Shenfield reminds us that the Soviet Union has well-established programmes of its own and is likely to develop space-based systems in a very different manner to those of the USA. If the USSR not only 'off-sets' but also 'emulates' the SDI's technologies it will be to support radically different military roles. Of crucial importance here is the potential development of 'space-to-earth weapons' (STEW) in an offensive role. Even though a new-generation Soviet ASAT is likely to be land based this kind of space strike weapon will clearly have an antisatellite capability. But more importantly STEW could give the USSR a weapon that is 'instantaneous' and against which existing devices for early warning would be useless. The potential of such weapons for strategic pre-emption and political 'decapitation' would be considerable and would deny the implementation of America's countervailing nuclear strategy. In addition, STEW could give to the USSR a genuine global military presence—an issue given much prominence in the Soviet literature but little discussed in the Western analyses of Soviet thinking and research.

Another technology driven by and distantly related to the original goals of SDI is that for terminal or point defence. However, its most likely near-term deployment is not as the fourth layer of an SDI system but as a single- or at best double-layer ATBM in Europe. It is uncertain the extent to which such a system would rely upon space-based sensor and control mechanisms, but it is undoubtedly the case that the SDI has led to an increased investment in and promotion of weapons that have an ATBM capability. It is also clear that the pressures that have driven the European members of NATO to consider an ATBM deployment are more to do with the domestic politics of maintaining support for the SDI in America than they are to do with the Soviet military threat in Europe. The primary importance of a European ATBM to the managers of SDI is as an intermediate deployment *en route* to a defence capability for the continental United States. It is clearly necessary to deal with the threat of the Warsaw Pact's fast-improving air forces and the recent deployments of accurate short-range dual-capable ballistic missiles. But it is equally important that this theatre problem should be countered by political or military means that are themselves tailored to the theatre rather than addressed in the name of a quixotic attempt to rid the world of nuclear weapons.

The SDI has nevertheless had the beneficial effect of forcing the superpowers and their allies to confront the implications of the militarisation of space and to assess which developments will serve their respective interests. That assessment begins from the strengths and weaknesses of the present strategic relationship. It is far more stable than the picture presented by the Reagan Administration in the early 1980s. It is not America's land-based ICBMs that are especially vulnerable but the command and control mechanisms that are required for the countervailing strategy. Whilst this may call into question America's, and by implication the USSR's, capacity for disciplined nuclear war fighting it leaves intact the capacity to retaliate at an unacceptable level.

With current levels of strategic warheads, and given the myriad problems of achieving an BMD with even a 95 per cent effectiveness, there is no prospect of a transition to a defensive strategic relationship. However, the testing or introduction of missile defences beyond the limits set by the 1972 ABM Treaty will not only destroy that arms control regime but will also push the superpowers to exceed the unratified SALT II limits on offensive strategic weapons. The United States has now commissioned the 131st B-52 bomber equipped with air-launched Cruise missiles thereby exceeding the limits, and the USSR has threatened that any attempt to deploy the SDI will lead to an increase in Soviet strategic offensive forces.

This warning notwithstanding, there has recently been a push by the Reagan Administration for an early deployment of 'phase one' of the SDI. The first indication of this came in late December 1986 when the DoD submitted a supplemental budget request of $110 million to the Congress for the research and development of a heavy lift space vehicle capable of taking large components into space. The request was made at the same time that the George C. Marshall Institute published a report recommending an early deployment of elements of the SDI. John Pike assessed that the supplemental budget request indicated that the Pentagon was considering an initial deployment in the 1990s.[2] This proved to be accurate and on several occasions throughout the first quarter of 1987 Casper Weinberger and congressional supporters, like Representative Jack Kemp and Senator Malcolm Wallop, called for the deployment of a first phase of the SDI 'as quickly as possible'. Weinberger envisages such a deployment to be the cornerstone of a nationwide shield, and part of the effort to produce a 'system that would protect the continent'.[3]

Throughout late January and early February 1987 there was dissension within the Administration not only about the early deployment but also about the related issue of how the 1972 ABM Treaty should be interpreted and whether or not it allows the testing of certain SDI components. Weinberger argued for a broad interpretation of the Treaty and claimed that deployment is possible by 1993–94, while George Shultz argued that a restrictive interpretation of the Treaty should be accepted and that the decision on SDI deployment could not be taken 'this year or next', thereby attempting to shift responsibility for a deployment decision onto the next Administration. Despite such confusion within the Administration, an elaborate plan for a first phase SDI to be deployed in the mid 1990s was announced by the DoD in February 1987.

The Pentagon plan consists of three levels of defence against incoming Soviet missiles and warheads. The first layer, designated the Space-Based Kinetic Kill Vehicle (SBKKV), consists of small rockets launched from space platforms that would destroy Soviet missiles during their boost phase. The second, the Exoatmospheric Reentry Interception System (ERIS) which is a development of the HOE programme, comprises ground-based missiles launched from the US that are intended to destroy warheads in the mid-course phase using a kinetic kill mechanism. A third layer, the High Endoatmospheric Defense Interceptor (HEDI), is another kinetic kill device that would also be launched from ground stations in the US for terminal and point defence. However, Alan Sherer, the HEDI programme manager, has said that 'We are still at the ground-test stage, so the technology is not yet proven.'[4] Despite this most of the technology envisaged for use in this plan is based upon known principles and marks a significant shift away from exotic technologies, such as directed energy and particle beams, that were promoted as the backbone of the SDI programme as recently as mid-1986.

The plan has run into much opposition not least because it is clear that the DoD has significantly reduced the operational requirements of each layer of the system so that it may be deployed at short notice. And the prestigious American Physical Society has recently published a study, based upon access to classified SDIO papers and reports, that concludes that the relevant technologies are so poorly developed that no informed decision about the deployment of an SDI programme can be made for at least ten years.[5] Added to this has been Senator Sam Nunn's very influential defence of the traditional and narrow interpretation of the 1972 ABM Treaty that has helped to crystallise congressional scepticism about the SDI and has made the testing of SDI components more difficult. Nevertheless, the DoD is currently calling for a series of Tiered Hierarchy Overlayed Research (THOR) tests to refine kinetic kill mechanisms for the interception of missiles and warheads in all stages of their flight. In the face of all the scientific opposition it is difficult to resist the conclusion of a *New York Times* leader that, 'The primary goal seems political: getting production lines running before Mr Reagan leaves office. That goal collides with any kind of orderly technical progress.'[6]

The combination of the growing and one-sided dependence

of the USA on satellites and the boost given to antisatellite technologies by SDI research will also undermine the tacit arms control that has very effectively inhibited ASAT deployment so far. If antisatellite systems are deployed they will threaten the effectiveness of any future space-based BMD, the offensive and retaliatory use of ICBMs, and the satellite assets used for communication, early-warning and arms control verification. Yet it is the USA that currently has the most reservations about opening negotiations for a treaty to ban or control ASAT deployments. It is disturbing that these reservations are linked to the desire to permit a wide range of BMD experiments in the short term, despite the fact, pointed to in several of the preceding chapters, that any weakening of arms control restraints will do more harm to the long-term strategic interests of the USA and NATO than to those of the USSR and the Warsaw Pact.

Many of the SDI's 'research' projects will continue, but the superpowers and their allies should acknowledge that the programme's original goals have long been abandoned as unattainable and undesirable. With the demise of these goals there is no longer the monolithic justification for the development of the vast array of technologies gathered under the SDI umbrella. They cannot, even in concert, achieve a transition to a new and more stable strategic order. Each technological thread should now be separated out of the tangled skein of the SDI programme and examined separately. From this perspective it will also be easier to make a realistic assessment of the various Soviet programmes and less easy for the USSR to justify its own exotic technology by reference to the 'threat' of the SDI. The new space and space-related technologies of the USA and the USSR will be developed in a world still dominated by offensive strategic weapons and retaliatory deterrence, and their capacity to stabilise or undermine the superpower relationship must be assessed in that context. The inevitability of technological progress does not impair the political volition of states to make its application a matter of choice and negotiation. Technology decrees that space will increasingly be militarised; political will may yet ensure that it is stabilised too.

NOTES

1. Dr Richard Garwin of Columbia University made several speeches during the summer of 1986 claiming that the Reagan Administration had used 'lies and distortions' to justify the SDI programme. Most of the SDI programmes he described as for C^3I and third-generation nuclear weapons which were justified solely as a hedge against a Soviet 'breakout' in advanced technology. According to Dr Garwin the 'US does not know what it is doing in Star Wars'.
2. John H. Cushman Jr. 'Pentagon seeks new SDI funds', *International Herald Tribune*, 1 January 1987, p. 1.
3. See, for example, Casper Weinberger, DoD press conference, Monday 12 January 1987, and John H. Cushman Jr. 'Weinburger urges early SDI deployment' *International Herald Tribune*, 14 January 1987, p. 2.
4. Alan Sherer's reservations and the DoD plan for ERIS and HEDI are reported in David E. Sanger, 'Skepticism mounts on deploying SDI', *International Herald Tribune*, 12 February 1987, pp. 1 and 6.
5. The report was prepared by a study group of 15 scientists under the co-chairmanship of Dr Nicolaas Bloembergen of Harvard University, and Dr C. K. N. Patel, Director of Research of AT&T Laboratories, and issued 23 April 1987.
6. *New York Times*, 11 March 1987.

Index